Reservoir Conformance Improvement

Reservoir Conformance Improvement

Robert D. Sydansk
Consultant

Laura Romero-Zerón
University of New Brunswick

Society of Petroleum Engineers

ISBN 978-1-55563-302-8

11 12 13 14 15 16 17 / 9 8 7 6 5 4 3 2 1

Society of Petroleum Engineers
222 Palisades Creek Drive
Richardson, TX 75080-2040 USA

http://store.spe.org
books@spe.org
1.972.952.9393

Preface

Identification, prediction, and understanding of conformance problems are vital in optimal design of improved and enhanced oil-recovery (EOR) processes. This book brings two topics together—conformance problems and conformance improvement techniques. Correct application of conformance improvement techniques can potentially lead to greatly improved oil-recovery (IOR) profitably from conventional oil reservoirs.

This book is practical rather than theoretical and reviews a number of subjects of wide interest. It provides an introduction to conformance, conformance problems, and conformance improvement for students, engineers, operation supervisors, and managers. The book incorporates material that is relevant to practitioners in the various areas of IOR and EOR, who are responsible for the implementation of conformance improvement techniques during the different stages of oil recovery. Furthermore, the book provides discussion of the importance of correctly identifying the specific conformance problems *before* the execution of any conformance-improvement technique. This might be the case with high-permeability thief zones where the injected water breaks through much earlier than expected (i.e., hours and days instead of weeks). In this case, the book offers guidelines to identify the nature of the high-permeability channeling so that the appropriate conformance improvement technique can be applied.

The material in this book reflects the authors' many years of experience in the industry and several decades of research, development, and field implementation of conformance improvement techniques. The book borrows a limited amount of the technical material from R.D. Sydansk's chapter, "Polymer, Gels, Foams, and Resins" in the *Petroleum Engineering Handbook* (2007).

We hope the readers find this book valuable in their pursuit of maximizing oil recovery from their oil reservoirs and fields.

Acknowledgments

Professor Hossein Kazemi of the Colorado School of Mines was the inspiration for and the guiding light during the writing of this book. The pioneering conformance improvement work of Randy Seright of New Mexico Tech is the foundation for a number of conformance improvement concepts discussed in the book. Randy is also acknowledged for providing a very helpful and detailed review of the initial manuscript version of the book. Professor Larry Lake of the University of Texas is acknowledged for an enlightening series of email exchanges with R.D. Sydansk on several key issues relating to conformance and conformance improvement; Larry provided some key technical insights that are the underpinnings for some of the material discussed in this book. Dr. Paul Willhite, Dr. M.R. Fassihi, and Dr. Gordon R. Moore are acknowledged for their many helpful suggestions that have facilitated to shape this book.

In Memoriam:
Robert Dunn Sydansk

This book is dedicated to the memory of my great friend Robert Dunn Sydansk. Bob showed extraordinary dedication toward advancing the science, engineering, and application of conformance improvement in reservoirs using polymers and gels. Bob invented the Cr(III)-carboxylate-HPAM gel conformance improvement technology. Weekly (and often daily) technical discussions with him were among the best parts of my job. I miss his insights, vision, and friendship.

Randall S. Seright
Senior Engineer, Research Group Head
New Mexico Petroleum Recovery Research Center

Contents

Chapter 1

Introduction

The intent of this book is multifaceted. The book provides a high-level and instructive overview of the petroleum engineering subjects of conformance, conformance problems, and conformance improvement during oil recovery. The book is *not* intended to be highly technical in nature or to be a detailed treatise.

The authors would like to raise the awareness level of, and interest in, the importance and the upside business and economic potential that is associated with conformance, conformance problems, and conformance improvement during oil-recovery operations. The book asserts that conformance improvement techniques are a possible and probable means to recover exceptionally large amounts of incremental oil and to create improved oil-recovery profitably from conventional oil reservoirs through the application of conformance improvement techniques and then, through subsequently applying, or continuing with, conventional and existing oil-recovery techniques. The book points out when and where the various conformance improvement techniques can be successfully applied. The authors attempt to demonstrate, especially to oil-company management, that conformance problems can possibly—or even probably— "kill" or greatly detract from evolving, proposed gas-injection and chemical-flooding enhanced-oil-recovery (EOR) processes if the conformance problems are not fully recognized and adequately addressed.

The book is hopefully written in an easy-to-read, engaging writing style and it is meant to provide the reader with pointers to more in-depth literature relating to the various conformance subject matters covered. The material covered in the book is intended to be as practical in nature as possible. At the end of the book are a glossary of terms and a list of abbreviations and nomenclature used in the book.

The intended audience of this book is varied and numerous. It begins with practicing reservoir and petroleum engineers and with oilfield production professionals—especially younger and newer engineers and professionals.

The book is also intended for petroleum-industry management and supervisors considering the application of improved oil recovery (IOR) and chemical-flooding or gas-flooding EOR to conventional oil reservoirs, who would like a better understanding and recognition of the potential for serious problems and pitfalls that conformance problems can and will probably pose to such EOR flooding. For instance, the performance of many of the chemical-flooding and gas-flooding EOR field-demonstration projects of the 1970s and 1980s were significantly affected by pervasive conformance problems that were not recognized at the time and, consequently, were poorly documented. Conformance during EOR is discussed in more detail in Chapter 11.

This book is also useful for graduate students who have an interest in petroleum-production conformance problems and mitigation techniques. The book is furthermore intended for any university-level student who would like an introduction to petroleum production conformance problems and the means to mitigate such problems.

An additional group for which this book becomes practical is anyone who has an interest in, or would like to learn or know more about, petroleum production conformance problems and their mitigation—and to do so on an overview and fundamental level. That is, persons who do not want an in-depth, lengthy, and detailed book on the subject. This includes individuals who do not want to deal with having to conduct an extensive literature search in order to obtain overview material on the subject of this book, and/or do not want to have to deal with extensive technical or mathematical jargon, notations, and concepts.

One final and important group for whom this book is intended is small, independent oil-producing operators who usually do not have significant technical staff to seek out, or develop, new, profitable oil-recovery technologies.

The focus of this book, especially within the last third of the core chapters of the book (Chapters 7–10), is improving conformance by operations or treatments that are applied within the petroleum reservoir itself. That is, during the last third of the book, the focus does *not* include conformance-improvement operations and techniques that involve mechanical means, which are applied directly within, or at, the wellbore, nor does the focus include conformance improvement operations and techniques that involve the drilling of new wells or operations that are conducted solely above ground. However, these latter types of conformance-improvement operations and techniques will be briefly reviewed, discussed, and listed in the earlier sections of the book.

The book is limited for the most part to conformance, conformance problems, and conformance improvement in *conventional oil reservoirs* that are not highly fractured in nature and reservoirs where the primary oil-recovery mechanism involves imbibition and gravity drainage. Conformance problems that exist in highly fractured reservoirs where the primary oil-recovery mechanism involves imbibition and gravity drainage are a special case of conformance problems. Such conformance problems are discussed in Chapter 3, Section 3.8.

Chapter 2

What Is Conformance? The Big Picture

2.1 Definition of Conformance

The term *conformance* in its truest and original form is defined as the measure of the volumetric sweep efficiency during an oil-recovery flood or process being conducted in an oil reservoir. As will be discussed further later in this book, the term conformance also refers, in certain instances and connotations, to a measure of, and the treatment of, excessive water production from petroleum reservoirs. Excessive water production often does impart *indirectly* a negative influence on the oil-recovery volumetric sweep efficiency during oil recovery within a given oil reservoir (Sydansk 2007).

The term *volumetric sweep efficiency* (E_V) represents the fraction (or percent) of pore volume in porous media that is swept by the injected fluid (Satter et al. 2008). Volumetric sweep efficiency is expressed by

$$E_V = E_A E_I \dotfill (2.1)$$

where E_A is the areal sweep efficiency, or the fraction of the pattern area that is swept by the displacing fluid, and E_I is the vertical sweep efficiency or the fraction of the pattern thickness that is swept by the displacing fluid (Satter et al. 2008).

For any given oil reservoir, the component of poor sweep efficiency that results from the oil-recovery drive fluid per se is aggravated as the viscosity of the drive fluid decreases. Within any reservoir having a given degree of permeability heterogeneity, as the viscosity of the drive/displacing fluid of an oil-recovery flooding operation decreases, the degree of viscous fingering, and the associated poor sweep efficiency, increases. The mathematical term that relates the viscosity of the oil-recovery drive fluid to conformance and sweep efficiency is the *mobility ratio*.

Mobility ratio, M, is defined as

$$M = \frac{\lambda_D}{\lambda_d} \dotfill (2.2)$$

where λ_D is the mobility of the oil-recovery *displacing* fluid phase and λ_d is the mobility of the *displaced* fluid phase. For a fixed volume of the oil-recovery drive fluid, areal and vertical sweep efficiency increases as *M* decreases. As applied within this book, the mobility of the displaced fluid phase is the mobility of the reservoir oil phase. Eq. 2.2 holds for a piston-like oil-recovery flooding operation in which the flood front is sharp. In such a case, mobility is defined as

$$\lambda_i = \frac{k_i}{\mu_i} \quad \ldots\ldots\ldots\ldots\ldots\ldots\ldots\ldots\ldots\ldots\ldots\ldots\ldots\ldots \quad (2.3)$$

where k_i is the relative permeability to phase *i* and μ_i is the viscosity of phase *i*. For a waterflood assuming piston-like flow, with only water flowing behind the front and only oil flowing ahead of the front, the relative permeabilities to water and oil are measured at residual-oil saturation and at interstitial water saturation, respectively. Thus, as the viscosity of the oil-recovery displacing/drive fluid is increased in a reservoir having a given degree of permeability heterogeneity, the sweep efficiency and the degree of the oil-recovery flood conformance are improved. Hence, *M* indicates the stability of a displacement process, with flow becoming unstable (nonuniform displacement front or viscous fingering) when *M* >1.0 (Sydansk 2007; Green and Willhite 1998).

Mobility issues and the associated mobility-induced viscous-fingering problems that occur during oil-recovery flooding operations result from the oil being significantly more viscous than the oil-recovery drive fluid (e.g., water).

2.2 Critical Premise

A critical premise of this book is that in order to recover mobile oil in a reservoir during an oil-recovery flooding operation, the mobile oil must be contacted and mobilized by the oil-recovery drive fluid. That is, the oil *must not be bypassed* by the oil-recovery drive fluid—at least in the time frame of interest. The critical premise is that if the mobile oil (in hard-to-reach locations) is not contacted by the oil-recovery fluid, such oil cannot be recovered in a timely (i.e., economic) fashion.

Of course, if the oil-recovery drive fluid (often immiscible) mobilizes a bank of oil, the oil-recovery drive fluid does not necessarily have to directly contact the oil that resides ahead of the contact point between the oil-recovery drive fluid and the oil bank when an oil bank is being mobilized. Another way of visualizing favorable conformance is that the oil-recovery drive-fluid front needs to flow nearly equally well, equally easily, and equally fast throughout the entire reservoir volume that is being flooding.

2.3 Illustration of Conformance and Generalized Conformance Problems

Fig. 2.1 can be used to visualize what conformance and conformance problems are about. In the upper portion of that figure, which is a vertical cutaway view of a matrix-rock oil reservoir between an injection and production well, if there was perfect *vertical conformance,* the flood front of the oil-recovery drive fluid (i.e., water) would move as a perfectly uniform vertical line from left to right in the figure (and within the reservoir). As actually depicted in the figure, there is a high-permeability channel midway vertically within the matrix-rock reservoir. This high-permeability channel causes poor vertical sweep efficiency; early breakthrough of the oil-recovery drive fluid (i.e., water); excessive and unnecessary oil-recovery drive-fluid (i.e., water) production; and undesirable and unnecessarily delayed oil production for the majority of the oil saturation that is residing in the lower-permeability portions of the matrix-rock reservoir located above and below the high permeability channel in Fig. 2.1.

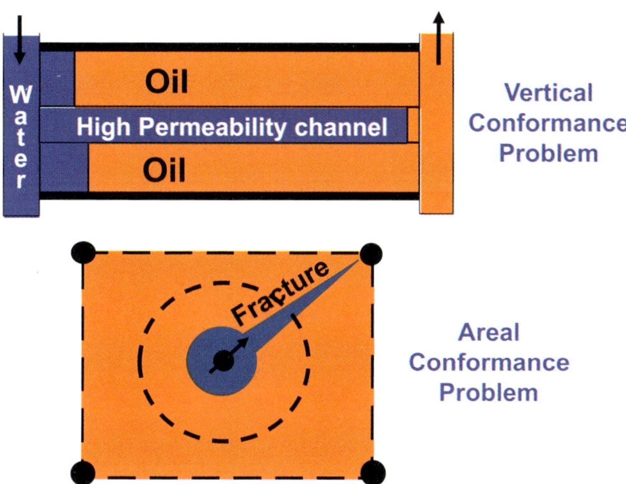

Fig. 2.1—Generalized matrix-rock and high-permeability-anomaly conformance problems.

The lower portion of Fig. 2.1 depicts an *areal conformance* problem. If there were perfect areal conformance, the flood front of the oil-recovery drive fluid (i.e., water) would expand outward continuously and radially with time from the injection well as a perfect circular flood front, indicated by the dashed line in the figure. In the areal-conformance-problem portion of the figure, there is drawn a vertical and high-permeability-anomaly (i.e., a fracture) that extends from the injection well to one of the four production wells. A single fracture extending directly from an injection well to a production well is highly unlikely, but is highly illustrative. However, the existence of a group of vertical fractures of a fracture network of intermediate density and having directional trends and fractures that extend from the vicinity of an injection well to the vicinity of a production well is quite common.

The single fracture in the lower portion of Fig. 2.1 will cause poor areal sweep efficiency; early breakthrough of the oil-recovery drive fluid (i.e., water); excessive and unnecessary oil-recovery drive-fluid (i.e., water) production; and undesirable and unnecessarily delayed oil production for the majority of the oil that is residing in the "bypassed" oil saturation that is located in the lower-permeability matrix-rock reservoir portion of the well pattern.

Vertical conformance problems quite often result from high-permeability strata residing within matrix-rock reservoirs. However, this is not always the case. For example, high-permeability-anomaly horizontal fractures can be present in relatively shallow oil reservoirs or high-permeability-anomaly horizontally-layered solution channels; or, interconnected vugular porosity can occur in carbonate reservoirs.

The mitigation and treatment of *vertical conformance problems* have been, at times, referred to as *profile modification treatments.*

Vertical conformance problems occurring in matrix-rock reservoirs normally involve the *radial flow* of fluids into, or out of, the wellbore for vertical wells when the wells are not hydraulically fractured.

Areal conformance problems often result from high-permeability-anomaly vertical fractures. But, this is not always the case. For example, areal conformance problems in matrix-rock reservoirs can be caused by large-scale high-permeability areal directional trends.

Flow of a reservoir fluid from the matrix-rock reservoir into a high-permeability anomaly (e.g., a fracture) usually involves some form of *linear flow*. An exception to this would be flow into a reservoir solution-channel tube.

If perfect conformance were to exist in a regular five-spot vertical well pattern during an oil-recovery flood, the flood front would reach all four of the offset producers at the same time and the flood front would reach the entire vertical interval of all four of the producing wells at the same time also. There never has been a reservoir that has exhibited perfect conformance during an oil-recovery flooding operation.

So, the questions are "How imperfect is the conformance for a given flooding operation in an oil reservoir?" and "What is the economic or other beneficial rate of return if a petroleum engineer were to implement a conformance-improvement operation or technique?"

Fig. 2.2 depicts a waterflooding operation being conducted in a heterogeneous matrix-rock reservoir in a five-spot well pattern, where the waterflood is experiencing both poor vertical and areal sweep. Thus, the waterflood in this well pattern is experiencing both vertical and areal conformance problems.

2.4 Conformance and the Rate of Oil Recovery

Conformance problems negatively affect the rate of oil recovery under any circumstance, but especially during oil-recovery flooding operations (e.g., waterflooding).

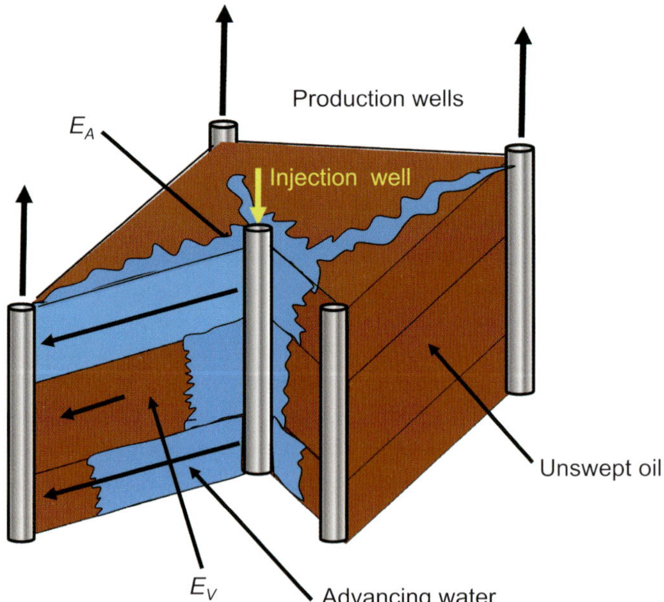

Fig. 2.2—A waterflood in a five-spot well pattern displaying both poor vertical and areal sweep efficiency (adapted from Cossé 1993).

The rate of oil recovery is a critical issue for oil industry business professionals because it affects how fast the investments for developing and depleting a given oil reservoir will be recouped. Thus, the rate of oil recovery indicates how great the profits for the investments in developing and depleting a given oil reservoir will be. The time value of money invested in developing and depleting a given oil reservoir is a central and crucial issue related to oilfield conformance issues.

The following discussion will consider the rate of oil recovery in terms of water-flooding a given reservoir. Similar arguments and considerations apply to any other oil-recovery flooding fluid or flooding operation. One of the first indications of serious reservoir conformance problems is an unexpectedly short time for water breakthrough during a waterflood. If the injected water breaks through much earlier than expected (e.g., breakthrough in hours, days, or a few weeks), then a serious conformance problem exists and there are likely problematic high-permeability anomalies residing in the reservoir and/or in the well pattern being flooded.

During waterflooding in a homogeneous reservoir, water and oil will be coproduced after water breakthrough occurs. Such coproduction of water and oil occurs under *fractional flow conditions* with increasing water/oil ratios (WORs) as a function of time.

When conformance problems exist, and after the conformance-problem water breakthrough, water and oil are also coproduced with often increasing WORs with time. In this case, coproduction of water and oil can go on for exceptionally long periods of time and do so at economic (often marginally economic) oil cuts. When conformance-problem water breakthrough occurs, the oil-recovery factor prior to any water production is often much less than expected (based on expected waterflood flow through homogeneous matrix rock only) and/or the coproduction oil with the water often does not tail off as fast as expected. The aforementioned are symptoms and indications of conformance problems.

Often conformance problems are the explanation of why so many reservoirs around the world, such as the reservoirs in the Big Horn basin of Wyoming, USA and the Permian basin of Texas and New Mexico, USA, have coproduced oil and water for so many years—often greater than 50 years. If water were just displacing oil in homogeneous reservoirs in these cases, reservoir engineering would dictate that the reservoir would be swept of oil to an economic WOR (e.g., 95% water cut) in a much shorter time frame.

Thus, conformance problems often greatly reduce the rate of oil recovery from a given oil reservoir, as compared to a similar oil reservoir that does not suffer from conformance problems.

2.5 Excessive Water Production as a Conformance Problem

Excessive water production due to conformance problems becomes an issue when it competes directly with oil production. This water usually flows to the wellbore through its own path, independent of the oil flow pathway. In such cases, a reduction of water production can often lead to a greater pressure drawdown and increase oil production rates (Seright et al. 2003).

Excessive water production as a conformance problem is discussed in more detail in Chapter 5.

2.6 The Broad Scope of Conformance Problems and Their Mitigation

Although this book focuses on conformance, conformance problems, and conformance improvement techniques as they relate to oil recovery from *conventional oil reservoirs*, it is crucial to note that the broad general aspects (discussed in this book) of conformance, conformance problems, and conformance improvement techniques are also applicable to conventional gas reservoirs, tight oil and gas reservoirs, and heavy oil reservoirs where the oil production occurs through wellbores rather than through heavy-oil, bitumen, and oil sands mining.

Chapter 3

Conformance: A Huge Potential for Improved Oil Recovery

3.1 Oil Recovery Term Definitions

Before discussing the huge potential for improved oil recovery (IOR) and enhanced oil recovery (EOR) resulting from conformance improvement considerations, it is important to review a number of definitions relating to the various forms of oil recovery.

Primary oil recovery involves oil production that is derived solely from the natural energy and forces within the reservoir (e.g., reservoir compaction due to overburden loading, expansion of dissolved gas in the oil, gas cap expansion, natural water drive, and gravity drainage). Reservoir-wide primary production occurs when there are only production wells in the field, and it does not involve any oil-recovery flooding operation. Primary oil recovery results either from naturally flowing wells or with the aid of artificial-lift production. Primary production has been informally referred to as the "first crop of oil production."

Secondary oil recovery refers to the use of "conventional" means to add energy to the reservoir to produce oil beyond primary oil recovery. Secondary recovery largely involves physical displacement processes to recover additional oil. Natural gas injection (for pressure maintenance) and waterflooding are the most widely applied secondary oil-recovery methods. Secondary production has also been referred to as the "second crop of oil production."

Tertiary oil recovery refers to the additional oil recovery that can be produced beyond primary and secondary oil recovery. During tertiary recovery, augmented-oil-flow and augmented-oil-recovery agents are injected into the reservoir to promote the tertiary oil recovery. These augmenting agents include (among others):

- Surfactants and/or alkaline agents to reduce oil/water interfacial tension in the porous reservoir rock and to increase the flood capillary number
- Thermal energy
- Miscible gas injection

Tertiary production has also been referred to as the "third crop of oil production." A possible shortcoming of the above definition of tertiary recovery is that some tertiary recovery techniques may be preferentially applied in the primary and secondary mode. For instance, thermal recovery methods (for reducing the viscosity of the oil) can be applied during secondary or tertiary recovery modes.

Tertiary oil recovery methods, for the most part, involve recovering oil that cannot be recovered economically by conventional waterflooding and pressure-maintenance gas flooding. Tertiary oil recovery often involves, at least in part, the recovery of residual oil saturation.

Conventional oil recovery includes all customary oil-recovery techniques, specifically primary and secondary oil-recovery methods. Conventional oil-recovery production excludes advanced oil-recovery methods and residual-saturation oil recovery (RSOR) methods. A shortcoming and the challenge in defining conventional oil recovery is that customary oil-recovery methods and techniques can vary somewhat with geographical locations around the world and with the evolution of time and recovery technologies.

Advanced oil recovery is a more generic term that includes all more technologically sophisticated oil-recovery techniques (e.g., infill drilling, horizontal wells, reservoir characterization and simulation, high-tech well completions, etc.). The term "advanced oil recovery" had a special connotation within the US Department of Energy in the early 1990s; however, it has been abandoned for the most part to avoid any confusion with the definitions of EOR and IOR (Stosur et al. 2003).

Incremental oil recovery is the additional recovery of oil that results from the application of any advanced, improved, or EOR operation or technique beyond the base oil production that was occurring before implementing the operation or technique.

EOR historically has been defined in several ways. The most widely accepted definition of EOR is oil recovery promoted by the injection of materials not normally present in an oil reservoir (Lake 1989). This definition of EOR covers an extremely broad range of materials that can be placed in a reservoir to improve oil recovery. This designation of EOR applies to all phases of oil recovery—including primary, secondary, and tertiary recovery. This broad meaning of EOR includes conformance improvement techniques and treatments in which conformance improvement fluids and materials are placed in an oil reservoir in order to produce incremental and improved oil recovery.

Some EOR practitioners have used more restrictive definitions of EOR. One of these more restrictive definitions was recovery of oil not recovered by secondary recovery (i.e., EOR equates to tertiary oil recovery). Another earlier and even much more restrictive meaning of EOR was an oil-recovery process for recovering the immobile residual oil saturation (defined below) within an oil reservoir. The term EOR also has been used to describe an advanced oil-recovery flooding operation that is conducted to recover tertiary oil production. However, in this book, the above first-listed, broader, more formal, and most accepted definition of EOR (Lake 1989) will be used.

IOR is any incremental oil recovery that results from the application of any EOR operation or advanced oil-recovery technique that is implemented during any type of ongoing oil-recovery process. Examples of IOR applications are any conformance improvement technique that is applied either during primary, secondary, or tertiary oil-recovery operations. Other examples of IOR applications are:

- Hydraulic fracturing-operations
- Scale-inhibition treatments
- Acid-stimulation procedures

IOR can also result from oilfield operations such as infill drilling and the use of horizontal wells. Historically, the terms EOR and IOR have been used interchangeably.

Mobile oil is the moveable oil saturation within a reservoir that can be mobilized and recovered during primary oil recovery and during a secondary oil-recovery flooding operation that uses an immiscible oil-recovery drive fluid (e.g. water). Mobile oil can also be mobilized and recovered by any tertiary oil-recovery flooding operation.

Residual oil saturation is the oil saturation in the matrix-rock reservoir (of any type of oil reservoir) that cannot be recovered by secondary oil-recovery flooding operations (e.g., waterflooding)—even in the presence of perfect flood conformance and/or flood sweep efficiency. Residual oil saturation cannot be recovered on a microscopic pore scale by an immiscible secondary oil-recovery flood and it is considered immobile residual oil saturation because such oil is held/trapped in the porous media by capillary forces (Green and Willhite 1998).

Residual-saturation oil recovery (*RSOR*) is a term that the authors of this book have coined for oil-recovery processes and technologies (e.g., surfactant or alkaline flooding) that are primarily intended to recover immobile residual oil saturation that remains (after primary and secondary recovery) in the reservoir due to capillary forces. An RSOR flooding technology can be applied to an oil reservoir that has been perfectly swept, with "perfect" conformance. However, in practice, RSOR floods are applied to reservoirs that have not been totally swept of their mobile oil saturation due to reservoir conformance issues and problems. As a result, in these instances, RSOR floods will also recover some of the mobile oil saturation still remaining within the reservoir.

Advanced Issues Discussion Box

Residual Oil Saturation and Mobile Oil

A more advanced discussion of *residual oil saturation* and *mobile oil* is presented here.

Residual oil saturation is synonymous with *irreducible oil saturation* and also is the *immobile oil saturation* during normal waterflooding. Residual oil saturation is normally *only part* of the oil saturation in the field setting, occurring at the economic end of a waterflood.

Residual oil saturation during waterflooding is most often determined in the laboratory setting by conducting flooding experiments in "representative" and "homogeneous" matrix-rock reservoir core material. The constant-differential-pressure flooding experiment is performed at reservoir conditions with reservoir fluids. The pressure gradient of the flooding experiments approximates the pressure gradients occurring in the bulk of the reservoir volume during the waterflooding operation. That is, residual oil saturation is determined in the laboratory setting in core plugs experiencing nearly perfect sweep efficiency. At these conditions, residual oil saturation is defined as the remaining oil saturation in the core plug when oil ceases to be produced (for core material that was initially at connate water saturation).

Capillary number, capillary forces, viscous forces, and capillarity are subjects related to both residual oil saturation and mobile oil saturation. If considering the mobilization and eventual recovery of oil drops or blobs existing at the pore scale in a *water-wet* matrix-rock, then the question of whether or not an oil drop can be mobilized and flow through the constrictive pore throat depends on the ratio of the viscous forces applied on the oil drop by the oil-recovery drive fluid (e.g., water), divided by the capillary forces acting on the oil drop (assuming gravity forces acting on the oil drop are negligible).

If the viscous forces are greater than the capillary forces, the oil drop will be mobilized and flow through the pore-throat constriction. On the other hand, if the viscous forces are less than the capillary forces, the oil drop will not be mobilized and will not flow through the pore-throat constriction (Green and Willhite 1998).

The *capillary number* for waterflooding, which is related to microscopic displacement and microscopic displacement efficiency, is mathematically defined as the viscous forces in play divided by the capillary forces operative on an oil drop in a water-wet pore. The capillary number is defined by the following equation:

$$N_c = \frac{\text{viscous forces}}{\text{capillary forces}} = \frac{v\mu_w}{\sigma_{ow}} \quad\dots\dots\dots\dots\dots\dots\dots\dots\dots\dots\dots\dots\dots\dots \quad (3.1)$$

where v is the interstitial velocity of the oil-recovery flooding fluid (i.e., water), μ_w is the viscosity of the oil-recovery flooding fluid, and σ_{ow} is the interfacial tension between the oil and the oil-recovery flooding fluid. This equation is not directly related to a measure of sweep efficiency or to the measure of conformance improvement but, rather, it is related to the recovery of residual oil saturation.

During waterflooding under normal oil-reservoir and normal waterflooding conditions, the capillary number is commonly $<10^{-6}$, and most commonly on the order of 10^{-7} (Green and Willhite 1998). At these low capillary numbers, "none" of the residual oil saturation (on average $S_{or} \approx 33\%$) will be recovered under normal waterflooding conditions. If the capillary number is increased up to a critical value that is specific to the particular reservoir and waterflooding conditions, the waterflood will recover *at least some* of the residual oil saturation. The amount of residual oil saturation recovered is a function of the magnitude of the capillary number of the flooding operation, as illustrated in **Fig. 3.1,** where RSOR stands for residual-saturation oil recovery and OOIP stands for original oil in place.

The waterflooding laboratory experiments are generally performed in "homogeneous" water-wet core plugs initially at connate water saturation under conditions involving a low flooding capillary number. At these conditions, approximately 67% of the mobile oil saturation in the core plugs is recoverable because the core rocks in these experiments are commonly homogeneous, and, as a result, the waterflood experienced nearly perfect sweep efficiency and no

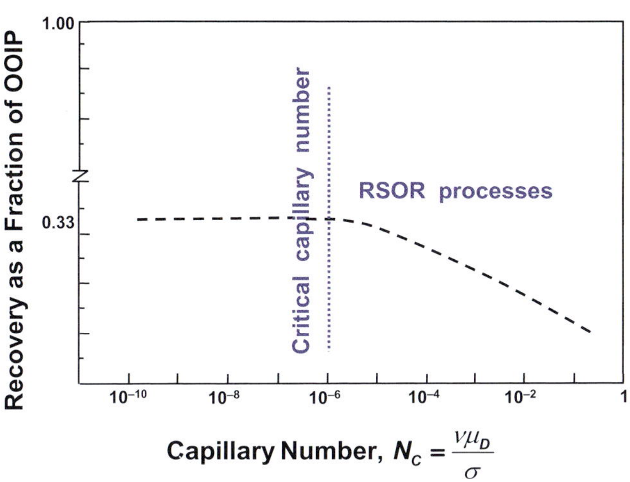

$$\text{Capillary Number, } N_C = \frac{v\mu_D}{\sigma}$$

Fig. 3.1—Oil recovery fraction vs. log N_c. Adapted from Taylor and Hawkins (1992).

conformance problems. Of course in an actual oil reservoir, a waterflood would invariably experience reservoir conformance problems, so that the entire mobile oil saturation in the reservoir would not be recoverable during a waterflood. Therefore, in the field setting, because of low flooding capillary numbers and conformance problems, any given waterflood will normally recover much less than the 67% of the OOIP such as was recovered in the laboratory using a homogeneous core plug during the flooding experiment of Fig. 3.1.

For waterflooding conducted at low flooding capillary numbers, the waterflood recovery factor will be, to the first approximation, independent of the flow rate of the water being flooded, and, thus, also of the pressure gradient of the waterflood.

For most waterflooding operations that are conducted in actual oil reservoirs, the capillary number of the flood is significantly low and most oilfield operators inject water at the maximum pressure possible without fracturing or parting the reservoir or causing only limited fracturing so as to avoid inducing flood conformance problems. If, alternatively, the viscosity of the oil-recovery displacement fluid is increased (in order to increase the numerator of the capillary number), then the flood injection rate and the associated flood pressure gradient will have to be reduced so as not to exceed the permitted maximum flood-injection pressure. However, reducing the flood injection rate is not economically desirable. As result, during common waterflooding operations there are no practical means to increase the numerator in the capillary number equation. Therefore, the only practical alternative to significantly increase the capillary number to recover residual oil saturation is to reduce the interfacial tension term in the denominator of the capillary number equation through the application of RSOR techniques (e.g., surfactants or alkalines).

Because capillary forces are responsible for trapping residual oil saturation in an oil reservoir during waterflooding operations, and because the application of

RSOR flooding technologies is required to recover any of the residual oil saturation, RSOR flooding can conceivably be termed a "capillary-trapped oil recovery" flooding technology.

Mobile oil (and mobile-oil saturation) is the oil saturation that can be mobilized and recovered by a secondary oil-recovery operation or by an immiscible oil-recovery drive fluid, such as waterflooding. Mobile oil and the mobile-oil saturation specifically exclude the residual oil saturation. The mobile-oil saturation of a reservoir specifically includes both the oil saturation that is recovered during conventional primary and secondary recovery operations and the significant mobile saturation that often remains in an oil reservoir at the end of normal (and economic) primary and secondary oil-recovery operations.

Bottom line: *Therefore, one-third (on the average) of the OOIP within oil reservoirs cannot be recovered by the application of any combination of primary/secondary oil-recovery methods and conformance-improvement techniques and operations.* Residual oil saturation can only be recovered by the application of RSOR techniques (e.g., techniques using surfactants and alkalines) that involve the reduction of the oil/water interfacial tension, therefore releasing the capillary-force that holds oil saturation "trapped" within matrix-rock reservoirs.

3.2 Focus on Conventional Oil Reservoirs

The primary focus of this book is on the application of conformance improvement operations and techniques to *conventional oil reservoirs.*

As discussed in a chapter of the *Petroleum Engineering Handbook* (Dusseault 2007), heavy crude oil is defined as liquid petroleum that has less than 20°API gravity or a viscosity higher than 200 cp at reservoir conditions. By default, and as used in this book, conventional oil reservoirs are reservoirs that contain—at reservoir conditions—crude oil having a viscosity of less than 200 cp or having API gravity greater than 20°. Conventional oils include crude oils referred to as "light" and "medium" oils. The term conventional oil, as used in this book, excludes all forms of heavy oil, bitumen, and oil from oil shale.

3.3 The One-Third Division of the Ultimate Oil-Recovery Fate of OOIP

One of the important messages that this book brings to the reader's attention is the *one-third division of the ultimate recovery fate of OOIP in conventional oil reservoirs* on a worldwide basis. As shown in **Fig. 3.2**, recovery of the OOIP on a worldwide average basis falls approximately into three equal thirds as follows:

1. One-third of the OOIP is *residual oil saturation* that can only be recovered (at least in part) by the application of RSOR techniques (Green and Willhite 1998).
2. One-third of the OOIP is *mobile oil saturation* that can and will be recovered by conventional primary and secondary oil-recovery techniques (Laherree 2003; Hirasaki et al. 2008).

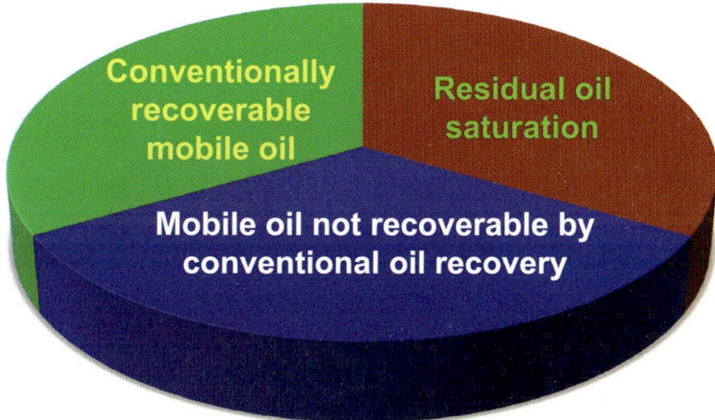

Fig. 3.2—The one-third ultimate oil-recovery fate of the average worldwide OOIP from conventional oil reservoirs.

3. One-third of the OOIP is *mobile oil saturation* that is not recovered by conventional primary and secondary oil-recovery techniques because of reservoir conformance problems.

It is important to note that the above one-third breakdown for the ultimate recovery fate for the OOIP within conventional oil reservoirs is *an average on a worldwide basis.*

The recovery factor of residual oil saturation (unrecoverable by conventional waterflooding and gas flooding) can range from <15% to >50% of the OOIP in various conventional oil reservoirs throughout the world.

The recovery factor for the mobile oil through the application of conventional primary and secondary oil-recovery methods can vary even much more widely throughout conventional oil reservoirs around the world—from <5% to >80% OOIP in various reservoirs.

Yet, the nearly equal one-third division of the ultimate oil-recovery fate (as shown in Fig. 3.2) for OOIP in conventional oil reservoirs does still hold presently on the worldwide average basis.

The assertion that only about one-third of the OOIP is presently recoverable by conventional primary and secondary oil-recovery techniques is supported, in part, by the following discussion.

Geffen (1973) stated that, for the US onshore oilfields (excluding Alaska), only 32.6% of the OOIP was expected to be recovered by conventional oil-recovery methods, as shown in **Fig. 3.3.** Thus, in turn, 67.4% of the OOIP can be targeted for EOR through a combination of conformance improvement and RSOR technologies.

A more recent report by Laherrere (2003) indicates that the *average oil-recovery factor for 4,000 fields outside the US and Canada is expected to be around 33% using conventional recovery methods.* These 4,000 oil fields were all over 10 million bbl in size and represented 1.3 trillion bbl—or 70%—of total world oil discovery. The data provided by Laherrere (2003) supports the assertion that, on average, only about one-third of the global oil reserves will be recovered by conventional primary and secondary oil-recovery techniques.

Similarly, the work of Lakatos and Lakatos-Szabó (2008) indicates that in the year 2000 the overall oil recovery efficiency from conventional oil reservoirs was

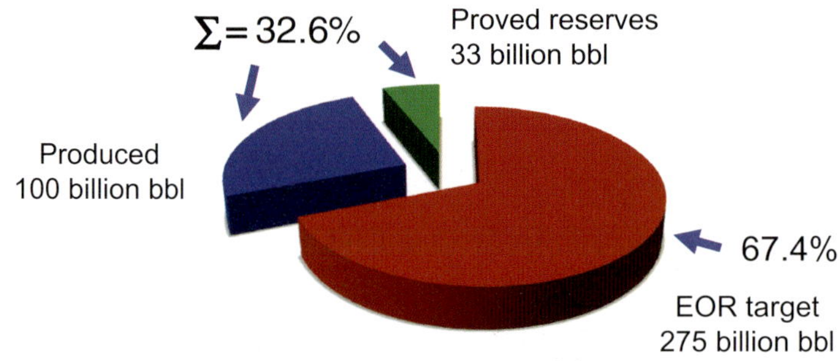

Fig. 3.3—Of the US onshore OOIP (excluding Alaska), only 32.6% is expected to be recovered by conventional oil-recovery methods (adapted from Geffen 1973).

33% OOIP. In this regard, Hirasaki et al. (2008) also stated: "It is generally considered that only about one third of petroleum present in known reservoirs is economically recoverable with established technology, i.e., primary recovery methods utilizing gas pressure and other natural forces in the reservoir and secondary recovery by waterflooding."

3.4 Encouraging Trends

There have been some encouraging, recent trends relating to "conventional-technique" oil-recovery factors and operations. There appears to be an increasing oil-recovery-factor trend for "conventional" oil-recovery operations conducted in the newer large fields that have been brought on production in Europe, the US, and scattered elsewhere in the world. This could increase in the future the average recovery factor of 33% for global conventional oil-recovery operations. For example, an impressive oil-recovery efficiency of 46% is reportedly being obtained for "conventional" oil-recovery operations in the North Sea (Sandrea and Sandrea 2007). The Statfjord oilfield in the Norwegian North Sea is reported to have attained an extraordinary 66% recovery efficiency during conventional production (Sandrea and Sandrea 2007).

Such encouraging conventional oil-recovery factors are likely attributed to some combination of sound reservoir management and newer oil-recovery technologies, such as conformance improvement by accurately targeting horizontal wells within the oil reservoirs in order to maximize mobile oil recovery. For instance, the Prudhoe Bay field in Alaska expects to recover 47% of the OOIP during its "conventional" oil-recovery phase (Sandrea and Sandrea 2007). It should be noted that miscible gas-injection was initiated "early on" at Prudhoe Bay.

In 2009, Saudi Aramco announced a corporate goal to increase the oil recovery rate by year 2029 to 70% (Petzet et al. 2009). The recovery rate of 70% would likely be achieved by a combination of conventional-oil-recovery, conformance improvement (e.g., exploitation of horizontal wells), and RSOR techniques.

Besides, in the future, conventional oil recovery may include more technologies and techniques than before (in the 20th century). For instance, horizontal well drilling might be considered a conventional oil-recovery technique, which previously was often considered to be an IOR technique and/or a conformance improvement technology.

3.5 A 2007 Breakdown of Estimated Global Oil Reserves

Sandrea and Sandrea (2007) reported the following breakdown analysis of the estimated recoverable global oil reserves:

- Total global conventional OOIP is 9,800 billion bbl
- Cumulative global conventional oil production (at year 2006) is 1,011 billion bbl
- Remaining global reserves, which are recoverable by conventional primary and secondary methods, are 1,147 billion bbl

There is likely some uncertainty in the accuracy of these global OOIP numbers. The true actual values of these numbers might reasonably be expected to be in variance with the reported numbers by a factor up to ±20%. However, for the sake of undertaking the following illustrative discussion, we will use these numbers and assume them to be accurate enough.

Table 3.1 provides a summary of the global oil reserves and associated breakdown analysis. As can be seen in the table, only 22% of the total global estimated OOIP is anticipated by Sandrea and Sandrea (2007) to be the world's ultimately recoverable reserves. This 22% average overall recovery factor is notably lower than the average 33% recovery factor reported by Laherrere (2003) and discussed in Section 3.3.

The information and estimations of worldwide oil recovery factors reported above from different sources concur on the same fact: a significant amount of unrecovered oil (>60% of the OOIP) is left behind in the reservoir after the application of conventional primary and secondary oil recovery methods. Therefore, there is great potential for the application of IOR and EOR technologies to advance oil recovery from mature reservoirs, particularly under the current worldwide energy needs.

3.6 Oil Reserves from Mature Reservoirs

One of the attractive features of pursuing the production of IOR and EOR in mature reservoirs by exploiting the application of conformance-improvement operations and techniques is that petroleum-producing professionals and oil companies know with relative certainty the exact location and volume of the target oil reserves in mature oil reservoirs. This is a much simpler endeavor than attempting to recover oil by exploration and the exploitation of new oil reservoirs. Whenever developing and bringing online a new oil reservoir, the operator often encounters a series of unexpected and less-than-helpful reservoir geological surprises that add cost and complexity to recovering the target volume of oil from the reservoir and/or that reduces the ultimate recovery factor from

<table>
<tr><td colspan="3" align="center">TABLE 3.1—BREAKDOWN OF GLOBAL CONVENTIONAL PROVED
RESERVES OF CRUDE OIL
[Worldwide conventional oil in 2006 (Sandrea and Sandrea 2007)]</td></tr>
<tr><td></td><td>Billion bbl of oil</td><td>% OOIP</td></tr>
<tr><td>Estimated OOIP</td><td>9,800</td><td>100</td></tr>
<tr><td>Cumulative production (through 2006)</td><td>1,011</td><td>10</td></tr>
<tr><td>World's remaining reserves</td><td>1,147</td><td>12</td></tr>
<tr><td>World's ultimately recoverable reserves</td><td>2,158</td><td>22</td></tr>
</table>

the reservoir through the application alone of conventional primary and secondary oil-recovery methods.

Oilfield operators who apply conformance-improvement IOR and EOR operations after having exploited conventional primary and secondary operations from a given oil reservoir, have an inherent advantage because they have acquired a relatively better understanding of the geological and other characteristics of the reservoir in question. The better the reservoir characterization available to the operator, the higher the probability of success, and the lower the risk, during the application of any oil-recovery operation that the operator might subsequently initiate in a given reservoir.

For economic reasons, when considering the recovery of the remaining oil (on average >60% of the OOIP) in mature reservoirs, it is, in general, often more advantageous to recover the conformance improvement EOR first because the implementation of conformance improvement EOR operations is often less expensive than the implementation of RSOR oil-recovery techniques. Furthermore, the cost of the EOR application can prove to be quite expensive if EOR flooding operations are conducted in the presence of substantial conformance problems, because of the resultant need of recycling the relatively costly EOR fluids through the reservoir.

3.7 Conformance-Improvement Operations Applied Within a Reservoir

Chapters 7 through 11 of this book focus on only those conformance-improvement operations and techniques that are applied within the oil reservoir itself. Therefore, these chapters will not cover conformance-improvement operations and techniques that:

- Are applied in, or at, the wellbore
- Involve drilling new wells and wellbores
- Are conducted solely above ground (e.g., pursuing the exploitation of better reservoir geological description and better reservoir computer simulation)

3.8 Highly Naturally Fractured Reservoirs

Highly naturally fractured conventional oil reservoirs (e.g., carbonate reservoirs), where the oil-recovery mechanisms are imbibition and gravity drainage, are considered non-classical conformance problems. These non-conventional conformance problems result because, for such highly fractured reservoirs, the oil-recovery drive fluid does not sweep and mobilize the oil directly from the individual matrix-rock reservoir blocks within the reservoir. In this case, the injected oil-recovery drive fluid (e.g., water) sweeps, mobilizes, and recovers oil that is produced (by imbibition and gravity drainage) into the natural fractures from the matrix.

However, in these reservoirs there are often unusually large and conductive natural fractures that contribute to conformance problems and poor oil recovery. These large fractures deter and/or prevent the injected oil-recovery fluid from uniformly and effectively sweeping and recovering (from the interwell fracture network) the oil that is produced from the matrix-rock reservoir blocks into the fracture network.

These highly conductive fractures can be appropriate candidates for the application of conformance-improvement methods and materials. Polymer gels have been used to successfully treat these particular types of conformance problems. In this situation, the polymer gels are used to improve waterflood sweep efficiency and conformance for flooding being conducted within the oil-reservoir fracture network itself. Additionally, the following conformance problems that can also occur in these reservoirs can be successfully treated with polymer gels to promote IOR:

- Casing leaks
- Flow behind pipe
- Coning through fractures
- Mobility-control viscous fingering (within the fractures)

Conformance problems and conformance treatments are discussed in the next three chapters.

Chapter 4

Conformance Problems

4.1 Root Cause

The root cause of petroleum-reservoir conformance problems is spatial variation in the fluid-flow capacity because of reservoir permeability heterogeneity. In addition, oil-recovery conformance problems can be exacerbated by, or dominated by, mobility-induced viscous fingering (Sydansk 2007). Mobility induced conformance issues and their resultant viscous fingering during oil-recovery flooding operations result from viscous oil being displaced by a relatively low-viscosity oil-recovery drive fluid (e.g., water).

In a highly homogeneous reservoir, it can become a challenging argument to determine whether the root cause of mobility induced conformance problems results from (a) miniscule variations in permeability within the reservoir, or (b) solely from the viscosity contrast between the "viscous" oil and the oil-recovery drive fluid (e.g., water).

Conformance issues resulting from vertical permeability heterogeneity can occur during vertical flow when the reservoir oil-recovery mechanism involves a bottom water drive or gas-cap expansion.

4.2 Types of Conformance Problems

In this chapter, typical oilfield conformance problems are briefly presented and discussed. The most common conformance issues are illustrated in **Fig. 4.1.** A brief review of the conformance problems depicted in Fig. 4.1 is provided—particularly, a concise analysis of how, and by what means, each of the conformance problems can be remedied (i.e., partially or fully mitigated). Similarly, a brief discussion is given on how difficult (relative to the other listed conformance problems) each of the problems is to remedy.

Later in this book, conformance-improvement methods, operations, techniques, and treatments are discussed and described in more detail (Chapters 6 through 10).

4.2.1 Mobility-Induced Viscous Fingering. Fig. 4.1a illustrates that mobility induced viscous fingering, even in a "homogeneous" reservoir, promotes and causes poor sweep efficiency (and conformance problems) during an oil-recovery flooding operation. The viscous fingering results from the displaced oil having a higher viscosity than

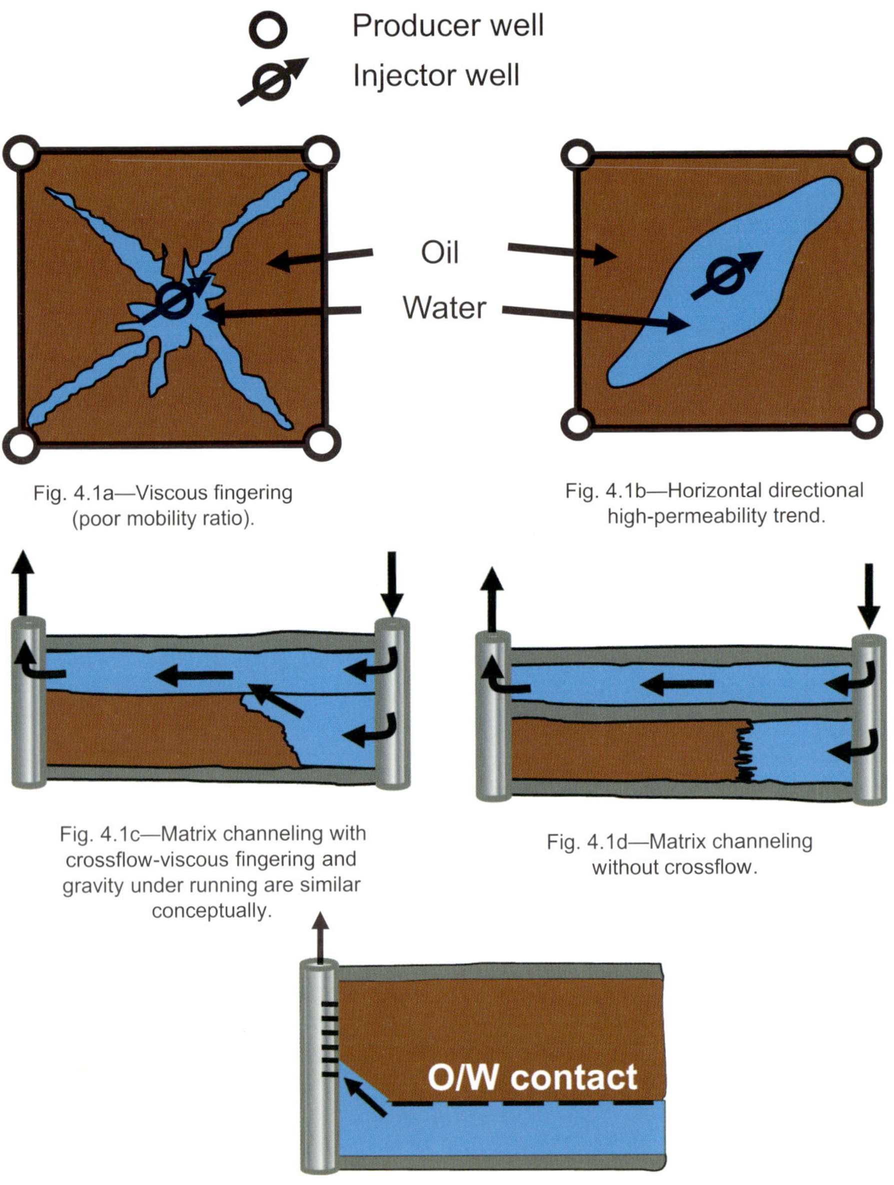

Fig. 4.1a—Viscous fingering (poor mobility ratio).

Fig. 4.1b—Horizontal directional high-permeability trend.

Fig. 4.1c—Matrix channeling with crossflow-viscous fingering and gravity under running are similar conceptually.

Fig. 4.1d—Matrix channeling without crossflow.

Fig. 4.1e—Matrix coning.

Fig. 4.1—Various conformance problems.

that of the oil-recovery-flood displacing fluid (e.g., water). This type of conformance problem is commonly found in reservoirs that contain relatively viscous oil (see Section 2.1). Mobility induced conformance issues can be, and often are, compounded and exacerbated by heterogeneous reservoir permeabilities.

Although not explicitly shown in Fig. 4.1, mobility induced viscous fingering can also occur in the vertical plane of a reservoir. Thus, mobility induced viscous fingering

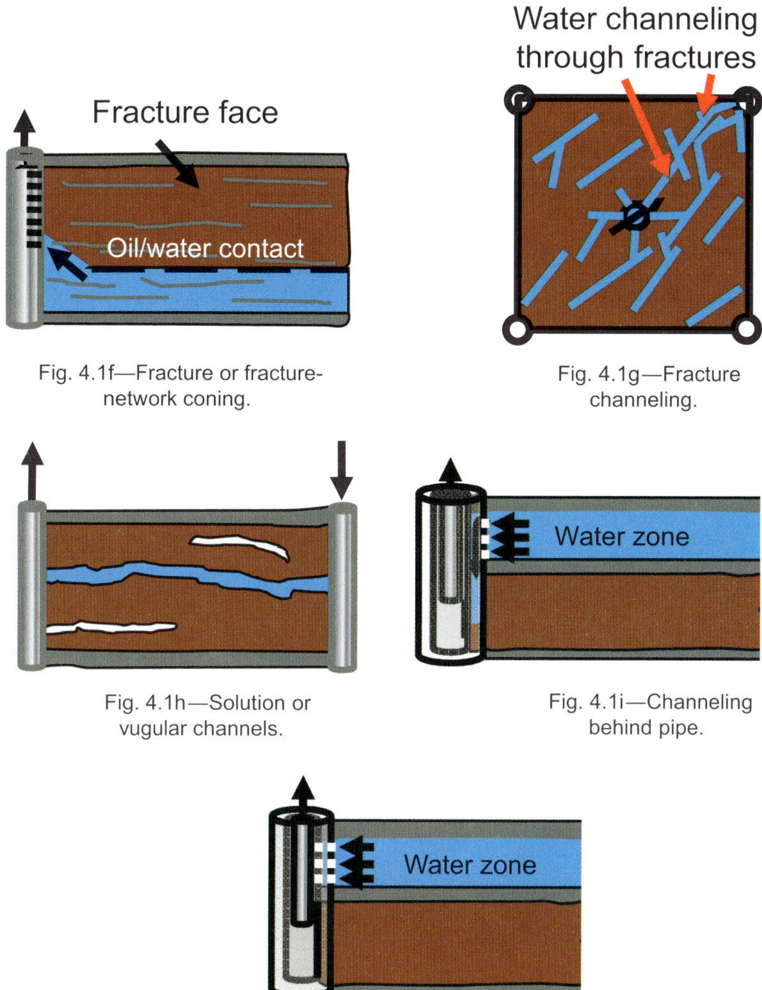

Fig. 4.1f—Fracture or fracture-network coning.

Fig. 4.1g—Fracture channeling.

Fig. 4.1h—Solution or vugular channels.

Fig. 4.1i—Channeling behind pipe.

Fig. 4.1j—Casing leaks.

Fig. 4.1 (continued)—Various conformance problems.

not only causes areal conformance problems as shown in Fig. 4.1a, but also can cause vertical conformance issues.

Viscous fingering induced by mobility-ratio problems can occur in the vertical flow direction when the reservoir oil-recovery mechanism involves a bottom water drive or gas-cap expansion.

The typically recommended (and easiest) conformance-improvement technique to overcome this conformance problem is to increase the viscosity of the oil-recovery drive fluid; for instance, through polymer flooding. For any conformance-improvement technique to remedy the mobility induced viscous fingering problem, the technique must not only function in the near-wellbore region, but also in the far-wellbore reservoir. Therefore, polymer flooding is the most widely applied conformance-improvement technology for increasing the viscosity of the drive fluid during oil-recovery flooding in a matrix-rock oil reservoir (Lake 1989; Sorbie 1991; Green and Willhite 1998; Sydansk 2007; Wang et al. 2008b).

4.2.2 High-Permeability Matrix-Rock Directional Trends. Fig. 4.1b illustrates the areal conformance problem and associated poor sweep efficiency that is caused by areal directional high-permeability trends.

This is a difficult conformance problem to remedy. If the goal is to mitigate this problem using the wells and the well patterns that are already in place, the normally recommended conformance-improvement technique is to increase the viscosity of the oil-recovery drive fluid (e.g., conduct a polymer flood). It is difficult to place a permeability-reducing conformance-improvement material like classical polymer gels or microgels far-wellbore enough in the reservoir (where its placement is required to treat this conformance problem) and to do so selectively, such that the permeability-reducing material is placed primarily within the high-permeability matrix-rock areal flow trends.

Therefore, the more reliable and more often employed means to mitigate this type of conformance problem is to (a) realign or strategically relocate vertical well locations areally within the oil field in question, or (b) drill and exploit strategically placed horizontal wells or other types of advanced wellbores.

4.2.3 High-Permeability Matrix-Rock Strata With Crossflow. Fig. 4.1c illustrates the vertical conformance problem that involves the existence of high-permeability matrix-rock strata, zones, or channels in matrix-rock reservoirs where *there is fluid crossflow and pressure communication* between the various reservoir strata. This conformance problem often results from geological strata of differing permeability overlaying one another in a petroleum reservoir. Oil-recovery drive fluids during flooding operations tend to preferentially channel and flow through the high-permeability strata.

This is a difficult conformance problem to remedy. The typically recommended conformance-improvement technique is to increase the viscosity of the oil-recovery drive fluid (e.g., conduct a polymer flood). In practice, it is often difficult to place a permeability-reducing conformance-improvement material like classical polymer gels or microgels far-wellbore enough in the reservoir (where its placement is required to effectively treat the problem) and to do so selectively such that the permeability-reducing material is placed primarily within the high-permeability matrix-rock geological strata.

Gravity Under Running. Although the oil-recovery conformance problem of *gravity under running* that can occur during waterflooding operations is fundamentally caused by gravity effects and not by permeability-heterogeneity problems. The single oil-recovery drive-fluid "finger" resulting from this problem occurring at the base of the oil reservoir can be mitigated by the same conformance-improvement tactics and techniques (e.g., polymer flooding) applied to mitigate the conformance problem of high-permeability matrix-rock strata with crossflow.

4.2.4 High-Permeability Matrix-Rock Strata Without Crossflow. Fig. 4.1d illustrates the vertical conformance problem that involves high-permeability matrix-rock strata, zones, or channels in matrix-rock reservoirs where there is *no fluid crossflow* *and pressure communication* between the various reservoir strata. For example, a continuous impermeable shale barrier/stratum could exist between the geological strata of differing permeabilities. This conformance problem often results from geological strata of differing permeability overlaying one another in a petroleum reservoir. Oil-recovery drive fluids during flooding operations tend to preferentially channel and flow through the high-permeability strata.

If an oilfield operator is unsure if the impermeable strata that exists between geological reservoir strata of differing permeabilities are continuous within the entire reservoir or within the well pattern of interest, it is normally recommended that the operator assume (and possibly err on the conservative side) that the impermeable strata are not necessarily continuous. There are techniques available (e.g., well testing) that can be used to determine if adjacent geological strata within an oil reservoir are in fluid-flow and pressure communication.

As compared to the apparently similar conformance problem discussed above in Section 4.2.3, the problem of high-permeability matrix-rock strata without crossflow is a relatively *easy problem to remedy*. There are a number of different conformance-improvement techniques that can be applied in this instance. One of the key essential elements needed to easily mitigate this problem is good quality and effective primary cement behind the casing over the entire well interval of interest.

A wide and diverse range of conformance-improvement techniques can be employed to remedy this type of conformance problem. First, strategic and appropriate well-completion techniques can be used to mitigate this difficulty. Second, a whole host of mechanical techniques can be used to mitigate this conformance issue such as sliding sleeves, packers and bridge plugs, straddle packers, and tubing patches. Third, a permeability-reducing agent (e.g., gels or resins) can be selectively placed in the offending high-permeability geological strata. In this case, where there is radial flow into or out of the wellbore, it is critical that no permeability-reducing treatment material be placed in the lower-permeability reservoir geological strata. This typically requires that the permeability-reducing agent be placed in the offending high-permeability strata using mechanical zone isolation (Seright 1988; Liang et al. 1993; Seright 2001). If a permeability-reducing agent is used to treat this problem, a "total" permeability shutoff material (for instance a gel or resin) is preferred. In this case, the permeability-reducing treatment normally does not need to be of a relatively large volume or placed very deep into the matrix-rock reservoir that is to be treated (often <10 ft radial penetration into the matrix-rock reservoir). Fourth, squeeze-cementing operations/techniques have historically been the most often applied technique to remedy this conformance problem. However, it is important to note that in numerous instances, the application of squeeze cementing to mitigate this conformance difficulty has not proven to be as successful and effective as desired, especially when treating production wells.

4.2.5 Water Coning Through Matrix Rock. Fig. 4.1e illustrates the vertical conformance problem of water coning up to the producing interval of a vertical well through matrix-rock reservoir. The water is coning up from an aquifer underlying the oil reservoir.

Gas coning down through matrix-rock reservoir (usually from an overlying gas cap) is a similar and analogous conformance problem to that of water coning up through matrix reservoir rock.

It is difficult, if not impossible, to implement a long-term solution (years up to the economic life of the well) for the coning problem when encountered in a vertical well. However, there are a number of techniques that will promote "short-term fixes," such as the placement of a disk-shaped permeability barrier radially away from a vertical production well at—or just below—the bottom of the producing interval to prevent water coning *for at least some time.* However, there are two important practical concerns with this approach. First, the horizontal-disk permeability barrier would have to be placed a great distance (many tens of feet) away from the production well. Second, if the permeability-reducing disk barrier is to involve the injection of a chemical fluid, such as a gelant (pre-gel) solution, the chemical fluid will not just propagate perfectly radially away from the wellbore over the vertical interval, as is required to be placed in a disk fashion. If injected at any substantial rate (as normally practiced in the oil-field), the injected chemical will flow not only in a horizontal radial fashion away from a vertical well, but it also will flow in the vertical direction in a "balloon" fashion. Consequently, the injected permeability-reducing chemical fluid will quickly invade upward into the producing geological interval, significantly reducing or shutting off oil production. Another quick and practical fix to the water coning problem is to decrease the fluid production rate. The practical long-term solution for stopping water coning is to plug the zone. If gelant is used to plug the offending zone, the hydrocarbon productive zones must be protected during gelant placement (Seright et al. 2008).

Also, an effective and favored long-term means to maximize oil production and to minimize undesirable water or gas coning is, again, the exploitation of horizontal wells.

Cusping Conformance Problems. Cusping conformance problems are somewhat similar to matrix-rock coning conformance problems. Cusping problems, like coning problems, are difficult, if not impossible, to treat so as to promote a long-term remedy (years up to the economic life of the producing well). Water cusping involves preferential water flow up through a high-permeability inclined reservoir strata (strata not in fluid-flow and pressure communication with adjacent oil-producing reservoir strata), where the reservoir strata inclines downward from the wellbore into an underlying aquifer. Gas cusping involves preferential gas flow down through a high-permeability inclined reservoir strata (strata not in fluid-flow and pressure communication with adjacent oil-producing reservoir strata), where the reservoir strata inclines upward from the wellbore into an overlying gas cap. As in the case of water conning, the application of gelants or gel treatments to remediate water and/or gas cusping has an extremely low probability of success when applied toward cusping occurring in unfractured matrix-rock reservoirs. The only long-term practical solution to stop water and/or gas cusping from the zone (other than decreasing the production rate) is to plug the zone. If gelant is used to plug the offending zone, the hydrocarbon productive zones in radial flow must be protected during gelant placement (Seright et al. 2008).

4.2.6 Water Coning Through Fractures. Fig. 4.1f illustrates the vertical conformance problem of water coning up through a vertical fracture or other high-permeability reservoir irregularity to the wellbore from an aquifer underlying the oil reservoir.

Gas coning down through fractures or other high-permeability anomaly (usually from an overlying gas cap) is a similar and analogous conformance problem to water coning up through fractures or other high-permeability anomaly.

In dramatic contrast to the conformance problem of water coning in matrix-rock reservoirs and involving vertical wells, water-coning conformance problems involving coning of water up through vertical fractures is a conformance problem that *can be easily and successfully treated,* at least for an *intermediate time frame and duration.* When using polymer-gels for water-shutoff-treatment applications, the excessive-water-production conformance problem that is most frequently, most successfully, and often profitably treated with polymer gels is water coning up to vertical wells through vertical fractures. However, it is important to note here that after successfully applying a polymer-gel treatment to block water coning in the near and intermediate distance from a vertical well and when producing at production rates significantly exceeding the critical coning rate (often the case), water will eventually cone up around the gel barrier that has been placed in the offending vertical fractures. In this case, the delay in further water coning can often be many months, or hopefully several years. When treating this problem with polymer gels, the performance and the effectiveness of the gel conformance-improvement treatment increases with increasing the volume of gel injected. Gel treatments for successfully treating water coning up through vertical fractures normally involve gel treatment volumes exceeding 1,000 bbl.

4.2.7 Fracture Channeling. Fig. 4.1g illustrates the areal conformance problem of fracture channeling of the oil-recovery drive fluid through a reservoir that possesses natural fractures. The number of channeling fractures can range from several to hundreds of natural fractures occurring within a single well pattern. When there are numerous fractures within a well pattern, all the fractures can be similarly aligned. However, there are often numerous fractures within a well pattern that tend to be a set of primary (highly conductive) fractures that are approximately orthogonal to a set of secondary (less conductive) fractures.

Fracture-channeling is a conformance problem (during oil-recovery flooding operations) that has, under appropriate circumstances, been successfully treated in an economically attractive and profitable manner through the application, for example, of injection-well polymer-gel sweep-improvement treatments. This is especially true when the conformance problem involves a fracture network of intermediate spatial density and a fracture network displaying areal directional trends. Polymer-gel conformance-improvement treatments for treating channeling natural-fracture networks in the intermediate and far-wellbore environment usually entail treatment volumes in the range of several thousand to multiple ten-thousand barrels of gel injected into the offending fractures.

4.2.8 Solution Channels and Interconnected Vuggy Porosity. Fig. 4.1h illustrates an areal and/or vertical conformance problem that can occur in carbonate reservoirs. This problem is caused by *solution channels or interconnected vuggy porosity.* Although these are two distinctive and different reservoir high-permeability anomalies, they are often lumped together because they both tend to occur in similar types of carbonate reservoirs and, at times, exist simultaneously in a single reservoir. Often, an operator has trouble distinguishing if the conformance problem is actually resulting from solution channels or interconnected vuggy porosity. Both solution channels and interconnected vuggy channels tend to have larger aperture flow channels than other common reservoir high-permeability anomalies (e.g., fractures).

Solution channels are distinct and nearly horizontal tube-like flow channels (normally of limited areal length) that occur under certain geological conditions in carbonate reservoirs, where these solution channels often have a diameter greater than approximately 0.5 mm. Solution channels only occur naturally in carbonate reservoirs, and the diameters of these natural solution channels are controlled by geological forces and factors.

CO_2 Flooding Considerations. Solution channels that are found in carbonate reservoirs that are being flooded with CO_2 are a special case in which they can be an especially challenging problem. Considering that CO_2 is acidic, and that the CO_2 will preferentially flow through the solution channels, CO_2 will act as a dissolving fluid that tends to enlarge the diameter of the carbonate solution channels. It may even be possible for the injected CO_2 to induce solution channels in the "matrix" rock of carbonate reservoirs—especially near-wellbore to the injection well. Near a CO_2 injection well in a carbonate reservoir, the diameters of solution channels can be eroded out to exceptionally large diameters (e.g., many centimeters in diameter).

Vugs. Vugs are naturally occurring (geologically induced) void spaces that exist under certain geological conditions in carbonate-reservoir matrix rock. The size of a vug can range from approximately 1 mm in diameter up to several centimeters in diameter. If the diameter of the vuggy void space becomes too large, the void space is then classified as a reservoir *cavern*. The exact diameter of the reservoir void space that must be exceeded in order for the void space to be characterized as a cavern, and not a vug, is not totally agreed upon.

Vugs in a carbonate reservoir that exist in isolation and do not connect with other vugs do not contribute to conformance problems. However, when the vug density becomes large enough and the vugs become interconnected, then such interconnected vuggy porosity becomes a reservoir high-permeability irregularity. Interconnected vuggy porosity (that contributes to conformance problems) often occurs as distinct and limited vertical strata within carbonate reservoirs, but may also have directional areal trends.

Like solutions channels in carbonate reservoirs, interconnected vuggy porosity can posses exceptionally high permeabilities and fluid-flow capacities—as compared to other common high-permeability anomalies within reservoirs (e.g., natural fractures).

Solution channels and interconnected vuggy porosity are oilfield conformance problems that can be, and have been, in certain and somewhat numerous instances, successfully treated in a similar fashion as described for treating the fracture-channeling conformance problem; that is, through the application of polymer gels. However, the challenge with the solution-channel and interconnected-vuggy-porosity conformance problems is that flow-channel apertures/widths of the high-permeability-anomalies can often be larger than those of the fracture types that have been successfully treated with polymer gels. At times, when encountering exceptionally large CO_2-flooding dissolution-enhanced solution channels in carbonate reservoirs, the use of materials such as foamed cement have been easily and successfully implemented for this purpose in substitution for, or in combination with, polymer gels.

4.2.9 Water Channeling Behind Pipe. Fig. 4.1i illustrates the excessive-water-production problem of water channeling behind pipe. In this case, the problematic water production is channeling from a water-bearing geological formation/stratum, down or up through some kind of channel that exists between the wellbore casing and the rock surface of the oil reservoir. Such water channeling can occur, for example, through:

- Continuous microannuli that exist either between the wellbore primary cement and the formation rock surface or between the primary cement and the casing pipe
- Imperfections or partial failures within the wellbore's primary cement job
- Partial or total absence of primary cement behind the casing over the vertical interval between the wellbore producing interval and the problematic water-producing strata above or below the producing interval

The excessive-water-production conformance problem of water channeling behind pipe is a problem that can be easily treated and mitigated. Use of polymer gels is an example of one technology that can be applied for this purpose when the water flow channels involve behind-pipe flow apertures less than 1 mm (0.04 in.). If the water flow-channel aperture is greater than 1 mm (0.04 in.), and especially if there is no primary cement behind the casing pipe, squeeze cementing is another technology that can be used to remedy water channeling behind pipe.

4.2.10 Casing Leaks. Fig. 4.1j illustrates the excessive-water-production problem of casing leaks. In this case, water from a water-bearing geological stratum/zone (either above or below the producing geological strata) leaks through the wellbore casing due to, for example:

- Corrosion failures in the wellbore metal casing
- Thread leaks in the casing-pipe coupling joints

The casing-leak excessive-water-production conformance problem is one that, in concept, can be easily treated and mitigated, but in reality can prove to be challenging to successfully repair. A number of different conformance-improvement techniques can be used to successfully treat casing-leak problems. These techniques include:

- Mechanical means such as tubing patches and straddle packers
- Chemical means such as applying polymer-gel or resin treatments
- Squeeze cementing in certain instances

4.2.11 Reservoir Compartmentalization. Reservoir compartmentalization is a special and extreme form of a conformance problem.

Reservoir-compartmentalization conformance problems can be viewed as a conformance problem in which extreme reservoir permeability heterogeneity occurs such that certain continuous portions of the reservoir volume have, essentially, zero permeability. Reservoir-compartmentalization conformance problems usually must be addressed and treated through drilling additional and strategically placed vertical wells and/or exploiting horizontal well configurations.

4.2.12 Closing Comments About Types of Conformance Problems. The previous discussion of the various types of conformance problems is *not* intended to provide the reader with an all-inclusive list of oilfield conformance problems, but rather to provide and discuss briefly an illustrative list of conformance problems.

An important point to note from the previous sections on conformance problems and possible treatments is that there is a great variation in the ease, and in the expected propensity for a favorable outcome for the various and different conformance problems. An appropriate strategy is to treat and remedy first the easier and more treatable conformance problems before attempting to address the more difficult conformance problems.

In the context of oilfield water management, Flores et al. (2008) provide an interesting list and discussion of conformance-problem types that can cause excessive water production.

4.3 Diagnosis of Conformance Problems

4.3.1 Requirements for Successful Diagnoses. It is clear that the proper diagnosis and analysis of the conformance problem or anomaly within the reservoir or near a wellbore leads to a better selection of the conformance problem treatment, which results in a more successful control of the problem, as stated by Soliman et al. (2000), who reviewed over 900 actual conformance treatments and determined that a success ratio exceeding 90% was obtained when the proper diagnostic analyses and tools were used.

Conformance problems can be classified into two main categories: (1) near-wellbore problems and (2) reservoir-related problems (Chou et al. 1994; Jaipatke and Dalrymple 2010). The diagnostic evaluation of conformance problems must consider these two conformance problem categories to pinpoint the source of the problem that might exist in a given area. According to Jaipatke and Dalrymple (2010) the diagnostic evaluation of conformance problems consists of the following elements:

- Well testing and reservoir monitoring
- Well-testing methods and equipment
- Numerical modeling and simulations

Diagnostic Evaluation Procedures, Tools, and Strategies. It is important to emphasize that a key aspect for the successful diagnosis of any conformance problem is the integration of multiple evaluation procedures and tools. Information provided by the use of only one procedure could be misleading, as is the case of the exclusive use of water/oil ratio (WOR) diagnostic plots, which could be misinterpreted, and should not be used alone to diagnose the specific cause of a water production problem (Seright et al. 2003). **Table 4.1** outlines some of the diagnostic evaluation procedures and tools commonly used for the identification of conformance problems.

Also, as part of a strategy for attacking excess water production problems (see Section 5.3), Seright et al. (2003) advocate (1) starting the diagnostic process using information that you already have and (2) if additional diagnostic methods are needed, choose methods to confirm or rule out the easiest problems first (and, by extension, diagnose the most difficult problems last). This strategy can be applied to the diagnosing of most conformance problems.

4.3.2 Diagnosing an Excess-Water-Production Problem—An Example. In addition to the several diagnostic methods listed in Table 4.1, Seright et al. (2003) recommend a low-cost diagnostic method based on injectivity/productivity calculations that should be considered when diagnosing any excess-water-production problem.

Determining Radial or Linear Flow. A critical aspect of diagnosing most excess-water-production problems is deciding whether fluid flow around the wellbore is radial or linear. Linear flow is associated with flow behind pipe, through fractures and/or fracture-like features, while radial flow generally occurs in matrix-rock reservoir when these features are absent. Straightforward calculations using the Darcy equation for radial flow (Eq. 4.1) determines the type of flow taking place at any well.

$$\frac{q}{\Delta p} = \sum \frac{kh}{\left[141.2 \mu ln\left(\frac{r_e}{r_w} \right) \right]} \quad\dots\dots\dots\dots\dots\dots\dots\dots\dots\dots\dots\dots\dots \quad (4.1)$$

<table>
<tr><td colspan="3" align="center">TABLE 4.1—BREAKDOWN OF COMMON DIAGNOSTIC EVALUATION PROCEDURES AND TOOLS AVAILABLE FOR THE IDENTIFICATION OF CONFORMANCE PROBLEMS (Jaipatke and Dalrymple 2010; Soliman et al. 2000; Seright et al. 2003)</td></tr>
<tr><td>Diagnostic Evaluation</td><td>Testing and Analytical Tools</td><td>Information Obtained</td></tr>
<tr><td>Well Testing</td><td>Vertical Interference "Tests"
• Pulse tests
• Formation testers
• Multiple-well testing
• Pressure-transient analyses</td><td>• Reservoir properties, horizontal and vertical permeability, crossflow between strata
• Information on reservoir nonidealities that should be analyzed in conjunction with geological data; detection and characterization of fractures (volume, permeability, spacing between fractures, orientation)
• Proper reservoir description with regard to static and dynamic properties</td></tr>
<tr><td>Reservoir Monitoring</td><td>• Analysis of production data (recovery factors, WORs) assisted by diagnostic plots to validate the quality of the production data (Anderson et al. 2006); examination of well production profile (Lane and Sanders 1995)
• Analysis of well history (e.g., recompletions, well stimulation, major workovers) (Anderson et al. 2006)
• Integration of reservoir description and reservoir simulation with multiple-reflection seismic surveys</td><td>• Monitoring of current movement of fluid saturations in a reservoir and prediction of future fluid-saturations movement, which provide vital information for delaying or preventing an early water or gas breakthrough.</td></tr>
</table>

TABLE 4.1—BREAKDOWN OF COMMON DIAGNOSTIC EVALUATION PROCEDURES AND TOOLS AVAILABLE FOR THE IDENTIFICATION OF CONFORMANCE PROBLEMS (Jaipatke and Dalrymple 2010; Soliman et al. 2000; Seright et al. 2003) (Cont.)

Diagnostic Evaluation	Testing and Analytical Tools	Information Obtained
Well-Testing Methods and Equipment	**Tracer Surveys** • Radioisotopes • Fluorescent dyes • Water-soluble alcohols • Water-soluble salts	• Indicate directional flow trends • Identify rapid interwell communication and reservoir continuity • Estimate volumetric sweep • Delineate flow barriers • Compare flow and sweep patterns • Characterization of fractured reservoirs: location and direction of fracture channels, fracture volume, fracture conductivity • Estimate the effectiveness of remedial treatments
	Logging Tool Services • Openhole logs: caliper, gamma, spontaneous potential (SP), and magnetic-resonance imaging (MRI) • Cement-evaluation logs: cement-bond logging (CBL) and ultrasonic bond logs • Casing-evaluation logs: multiarm caliper tool, casing-inspection tool (CIT), flux-leakage/eddy -current (FL/EC) tool, circumferential acoustic scanning tool (CAST), and pulse-echo tool (PET) • Pulsed-neutron logs • Production logs: fluid-density tool, hydro tool, spinner tool, pressure tool, and temperature tool • Seismic methods	• Porosity, permeability, irreducible water saturation, fluid quantification (oil, water, and gas), water-cut prediction by integrating MRI log with resistivity logs. Reservoir heterogeneities. Identification of fractures or fracture-like features • Current condition of the cement annulus and diagnosis of potential fluid-flow paths • Integrity of the casing • Detection of channels outside the casing, leaking tubular, and water production • Crossflow between strata • Water influx, rate, and direction of flow
	Real-Time Downhole Video Services	• Identify wellbore problems, fluid turbulence, and flow direction. This information is useful to establish fluid migrations through the wellbore and into "thief" formations. Similarly, it allows planning reservoir and well treatments while in progress and confirms post-treatment well conditions
Numerical and Experimental Modeling and Simulations	• 3D, three-phase, four-component, pseudo-compositional, and non-isothermal coupled reservoir/wellbore simulators	• Identification and understanding of the conformance problem • Prediction of the effect of conformance-improvement treatments on reservoir performance • Prediction of maximum water-free production rates • Estimation of breakthrough time, water-cut performance, and/or economic production rate

where

q = total injection or production rate, (B/D);

Δp = pressure drawdown of buildup, (psi);

k = effective rock permeability, (md) (the permeability data should be taken from core analyses, log data, or from pressure transient analyses—it should not be taken from production data);

h = net pay, (ft);

μ = viscosity, (cp) [if the well is a water injector or is producing a very high water cut, then the viscosity of water can be used (at the appropriate temperature); if the oil cut is significant, there may be value in performing two calculations using Eq. 4.1—one using water viscosity and one using oil viscosity];

r_e = external drainage radius, (ft);

r_w = wellbore radius, (ft); and

$\ln(r_e/r_w)$ = the natural log term in Eq. 4.1 can be assumed to have a value of 6 or 7.

If the actual injectivity or productivity for a well (i.e., the left side of Eq. 4.1, $\dfrac{q}{\Delta p}$, in B/D/psi) is five or more times greater than the injectivity or productivity calculated with the Darcy equation for radial flow (i.e., the right side of Eq. 4.1), the well probably suffers from a linear flow problem (flow behind pipe, flow through fractures and/or fracture-like features). Thus, linear flow has prevailed when

$$\frac{q}{\Delta p} \gg \sum \frac{kh}{\left[141.2\mu ln\left(\frac{r_e}{r_w}\right)\right]} \quad\quad\quad\quad\quad\quad\quad\quad\quad\quad\quad (4.2)$$

and radial flow becomes likely when

$$\frac{q}{\Delta p} \leq \sum \frac{kh}{\left[141.2\mu ln\left(\frac{r_e}{r_w}\right)\right]} \quad\quad\quad\quad\quad\quad\quad\quad\quad\quad\quad (4.3)$$

Seright et al. (2003) emphasize the fact that in practical applications, uncertainty exists for a significant range of conditions that do not satisfy either Eq. 4.2 or Eq. 4.3, so that injectivity/productivity calculations will not always distinguish between radial and linear flow. However, this simple diagnosis procedure frequently provides a definitive indication of the flow geometry near the wellbore.

Crossflow Testing. If the conformance problem is identified as radial flow, then the possibility of crossflow between reservoir strata must be addressed. Several methods are available to determine if crossflow exists between strata as outlined in Table 4.1. The most straightforward method tests pressure differences between zones. A packer is commonly placed between two zones, and one of the zones is allowed to pressure up. If a significant pressure can be maintained across the packer, effective barriers to crossflow exist between zones. If a pressure difference cannot be maintained, crossflow between the zones can occur (Seright et al. 2003).

Excessive water production as a conformance problem is discussed in detail in the next chapter.

4.3.3 Tracer Testing Examples.
Tracer testing in the petroleum industry started in 1962 and at the present time is a well-established technique to evaluate interwell

communication as well as reservoir heterogeneity (Asadi and Shook 2010). Examples of water-soluble and gas-soluble tracer field applications are presented in the subsequent paragraphs.

A Water-Soluble Tracer Field Application. The first example summarizes the application of water-soluble chemical tracers in a field located in the Permian Basin (North America) as reported by Asadi and Shook (2010). The objective of this field tracer application was to identify the source of a significant increase in water cut from almost all of the producers in the pattern under evaluation. The section of the field selected for tracer testing included three injection wells and nine monitoring wells patterned randomly. Three different chemical tracers were selected for injection. A volumetric method, a low detection limit of 50 parts per trillion (ppt), and a safety factor for the uncertainties in reservoir parameters were used to calculate the amount of tracer needed for injection in each well. To assure detection of any high-permeability zones between the injection and production wells, an aggressive sampling schedule of one sample per day per well for the first four weeks was conducted for the monitoring wells. Selected samples were analyzed for tracer(s) detection. Additionally, one water sample per producer was collected one month before tracer injection for analyses to establish a baseline on the reservoir fluid. The three tracers were injected in a period of two days with each tracer being injected during a six-hour injection period to assure appropriate mixing within the rock formation.

The tracer surveys indicated the presence of a highly conductive flow path between the injectors and several of the production wells. These wells were, therefore, thought possibly to be connected via a major east-west natural fracture that extended to all producers by micro-fractures. It seemed that the high volume producers created pressure sinks that may have drawn injection fluid toward them. The knowledge gained from the tracer work allowed the company to adjust the waterflooding injection patterns, which improved sweep efficiency and, therefore, oil recovery. More details on this case history of field application of chemical tracers can be found in Asadi and Shook (2010).

A Gas-Soluble Tracer Field Application. A pioneer field case history of radioactive (gas-soluble) tracers and techniques in the Fairway (James Lime) reservoir was reported by Calhoun and Hurford in 1970. The objective of the radioactive gas tracers program in this field was to determine the source of gas breakthrough that was taking place during the alternate gas-water displacement process in a complex reservoir exhibiting various degrees of stratification. The radioactive tracers used were tritiated hydrogen (10 Ci), Krypton 85 (10 Ci), and tritiated methane (10 Ci). Phase I of the radioactive tracer program was initiated in August 1966, when Krypton 85 and tritiated hydrogen were injected into two different wells in the first line of gas injection wells. Periodic samples of the produced gas from offset wells was started shortly after the first tracer injections, but no tracers were detected in the producing wells until several months later. Phase II (December 1966) involved the second row of injectors and consisted of the injection of Krypton 85, tritiated hydrogen, and tritiated methane in three different wells. Phase III consisted of the injection of radioactive isotopes in three different wells of the third row of injection wells. The radioactive isotopes tracer program performed at the Fairway field demonstrated the source of gas breakthrough in approximately 25 wells. Continuous sampling of the wells after breakthrough indicated changing frontal configurations and fluid migration patterns. Tracer responses correlated with alterations in injection cycles. Fluid movements inferred from tracer measurements were consistent with variations in the reservoir pressure gradients as a result of the step-wise repressuring plan. Tracer responses also proved the need for controlled injection and withdrawals to even out sweep configurations. Furthermore, tracers indicated areas where injection rates and gas-water cycles should be changed to reduce fingering of the injected gas. For the Fairway field, this knowledge was extremely useful in alternating injection cycles and rates to control gas/oil ratios (GORs) at a reasonable level. Analysis of the relationship between tracer responses, producing GORs, and injection cycles indicated the need for smaller slugs and more frequent alternations of gas and water cycles. More specific details of this case history are provided in Calhoun and Hurford (1970).

Once the conformance problem or problems within the reservoir or near a wellbore have been properly identified, a recommended strategy and methodology to successfully apply a conformance treatment can be found later in this book in Section 5.3.

4.4 Specific Conformance Problems Require Distinct Conformance-Improvement Techniques

It is critically important to understand that specific oilfield conformance problems can be successfully treated and remedied (or partially remedied) only with distinct and certain types of conformance-improvement techniques and operations. Thus, one of the main challenges in successfully treating any given oilfield conformance problem is to correctly identify and/or deduce the nature of the conformance problem that is to be treated and then to choose from a limited number of conformance-improvement techniques or operations that can be used to successfully treat the identified reservoir conformance issue.

Therefore, it is crucial to the success of an oilfield conformance-improvement operation to determine if the offending conformance problem involves high-permeability anomaly flow channels (having permeabilities greater than 10 darcies) or if it involves high-permeability matrix-rock flow channels (having permeabilities less than 10 darcies) (Sydansk 2007). It is essential to customize the conformance-improvement treatment to the specific conformance problem; otherwise the treatment is often doomed to failure.

4.5 High-Permeability Anomalies

Although oil-reservoir high-permeability anomalies have been previously mentioned and briefly discussed in this book, this section discusses in more detail high-permeability irregularities. High-permeability-anomaly flow channels that occur within oil reservoirs are a critically important concept as it relates to a number of conformance problems occurring within an oil reservoir.

High-permeability-anomaly flow channels within oil reservoirs can be viewed and defined in one of two ways. First, they can be defined in terms of the absolute permeability of the flow channel; that is, a flow channel that has permeability greater than 10 darcies. Second, high-permeability-anomaly flow channels within petroleum reservoirs can be defined in terms of permeability contrast (say, by a factor greater than 1,000) between the permeability of the high-permeability anomaly and the permeability of the matrix-rock reservoir of the bulk of the petroleum reservoir (Sydansk 2007).

The second definition is more applicable when discussing high-permeability anomalies occurring in tight oil or gas reservoirs (e.g., microfractures). The first definition given above for a high-permeability anomaly is the definition used in this book.

The fluid-flow capacity of reservoir high-permeability-anomalies is only a part of, and an extension of, the fluid-flow-capacity continuum of fluid-flow paths that exist within oil reservoirs. A list of some (but not all) high-permeability anomalies that can occur within oil reservoirs is as follows:

- Fractures
- Faults and joints
- Solution channels
- Interconnected vuggy porosity
- Karsting
- Rubblized zones
- Cobble packs
- Unconsolidated coarse sand
- Multi-Darcy matrix reservoir rock

4.6 Fracture Conformance Problems

Reservoir fractures are the most common form of high-permeability anomalies occurring in oil reservoirs. Fractures, as used in the connotation of reservoir conformance problems, are cracks that occur in consolidated reservoir matrix rock. The existence of conformance-problems resulting from natural fractures occurring in carbonate reservoirs is very likely. The probability of natural fractures occurring in sandstone reservoirs diminishes because these reservoirs tend to become unconsolidated. Also, the probability of open natural fractures occurring in reservoirs diminishes for deep reservoirs. At reservoir depths greater than approximately 8,000 ft, the geological overburden pressure becomes great enough to often close up open natural fractures in petroleum reservoirs (Sydansk 2007).

At reservoir depths less than 2,000 ft, reservoir fractures tend to be horizontally orientated, potentially causing serious *vertical* conformance problems during oil-recovery flooding operations. At reservoir depths greater than 4,000 ft, reservoir fractures tend to be vertically orientated, potentially causing serious *areal* conformance problems during oil-recovery flooding operations (e.g., waterflooding) (Sydansk 2007). Natural vertical fractures in petroleum reservoirs can occur as single entities or as part of an interconnected fracture network.

There are two distinct types of single reservoir fractures. The first type of these single and isolated fractures is fractures that naturally occur in petroleum reservoirs. The second type of single reservoir fractures is hydraulic fractures that are artificially induced from either injection or production wells. Often, hydraulically induced fractures are "single" two-winged fractures that extend out in opposite directions from the well from which they were induced. However, some oil-reservoir hydraulic fractures can be branched in nature (and not simple single fractures). In general, hydraulic fractures are not as problematic conformance-wise as are certain types and sets of natural fractures. In fact, hydraulic fractures are purposefully induced in many oil reservoirs to aid in efficiently producing oil—in some cases oil that could not be produced otherwise.

4.6.1 Effect of Fracture Orientation on Conformance. The spatial orientation of vertical fractures in a petroleum reservoir can have a profound effect on whether these fractures will cause conformance problems.

Hydraulic fractures are no exception to this. **Fig. 4.2** depicts the effect that the orientation of a two-winged hydraulic fracture at injection wells can have on conformance and sweep efficiency during an oil-recovery flooding operation. The figure depicts a row of injection wells and a row of production wells within a field developed using a normal five-spot well pattern. When the hydraulic fracture orientation runs parallel to the line of injection wells, the hydraulic fractures actually help to improve the flood sweep efficiency and conformance by promoting better sweep efficiency of the oil-recovery drive fluid as it flows through the reservoir matrix rock to the producing wells. In contrast, when the orientation of the hydraulic fracture is in a direct line between an injection well and a producing well, and when the fracture extends a significant distance from the injection well to the production well, it promotes poor flood sweep efficiency and can be the cause of a serious well-pattern conformance problem.

4.6.2 Types of Fractures and Fracture Networks. There are several classes of natural fractures, as is depicted in **Fig. 4.3.** The first class is a single and/or isolated natural fracture located within a petroleum reservoir or within a well pattern. Whether such fractures contribute to conformance problems or prove to be benign in a conformance-problem sense very much depends on the fracture properties of these single and isolated fractures. Such fracture properties include:

- Fracture orientation
- Fracture location with respect to well location
- Fracture width
- Fracture length

If a given oil reservoir contains this class of natural fractures, the single fracture or set of isolated fractures will have to be considered and evaluated on a case-by-case basis

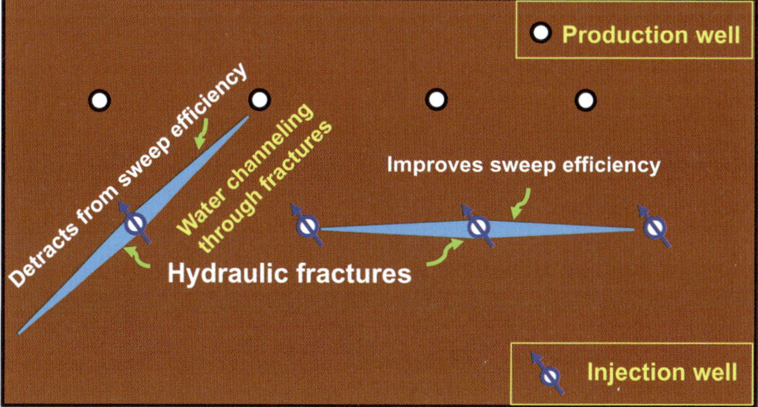

Fig. 4.2—Effect of the orientation of an injection-well hydraulic fracture on sweep efficiency.

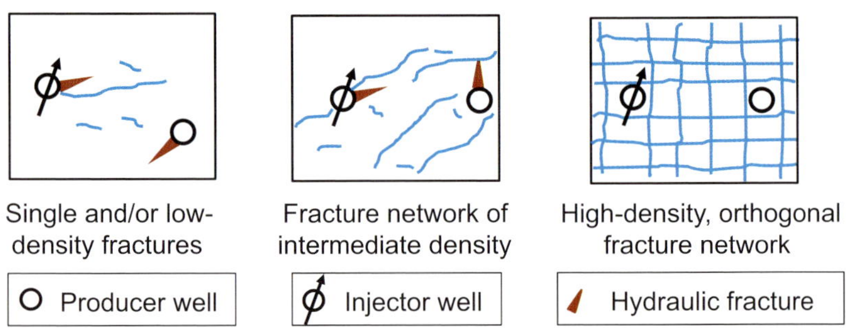

Fig. 4.3—Conceptual classes of reservoir vertical fractures and fracture networks.

to determine if they are causing any measurable conformance problems, and if so, what to do about it (Sydansk 2007).

The second class of natural fractures, as it relates to conformance problems, is a natural fracture network of intermediate fracture-spacing density as shown conceptually in Fig. 4.3. If such a fracture network tends to show directional permeability trends (as these fracture networks often do) the fracture network can cause serious areal conformance problems and poor areal sweep efficiency during oil-recovery flooding operations (e.g., waterflooding). Fortunately, this type of fracture conformance problem has proven, in certain instances, to be successfully treated through the application of conformance-improvement polymer-gel treatments (Sydansk 2007).

The third class of natural fractures is natural fracture networks possessing a high density of fractures. Fracture spacing can be as small as two feet or less. Such fracture networks often have orthogonally orientated sets of fractures. The primary oil-recovery mechanism in such reservoirs (often carbonate reservoirs) involves a combination of imbibition and gravity drainage. If the fracture sets of such highly fractured reservoirs are somewhat homogeneous in properties, then these naturally fractured reservoirs usually are not suitable candidates for applying conformance-improvement operations. In fact, when viewed from a relatively large macroscopic prospective, such reservoirs begin to approach homogeneity in reservoir properties.

However, it is often the case that in such high-density fracture networks, a relatively small number of the fractures have exceptionally large fracture apertures/widths and, therefore, possess exceptionally large permeabilities (fluid-flow capacities). The existence of such highly conductive fractures can result in poor sweep efficiency during oil-recovery flooding operations that are intended to sweep the imbibition-produced oil that is residing in the fractures to a producing well. When these anomalous highly conductive fractures do exist in a highly fractured reservoir that is undergoing a flooding operation, then a flooding-operation conformance problem exists. Such flooding conformance problems in highly fractured reservoirs and involving anomalous highly conductive fractures can be successfully treated with a conformance-improvement technique, such as the application of an appropriate polymer-gel treatment.

4.7 Effect of Crossflow on Conformance-Improvement Operations

In matrix-rock oil reservoirs suffering from vertical conformance problems caused by geological strata of differing permeability overlaying one another, a critically important question relevant to the successful application of conformance-improvement operations is whether or not vertical fluid-flow, crossflow, and pressure communication can occur between the various reservoir strata or zones (Sydansk 2007; Sorbie and Seright 1992).

If crossflow does not exist, then such conformance problems can easily be treated and mitigated by a number of different conformance-improvement techniques that can be applied either in the wellbore or in the reservoir near-wellbore environment and can be treated either from the injection-well or the production-well side. Two fluid-bearing reservoir geological strata will not be in vertical fluid-flow communication and in pressure communication with one another if, for example, an impermeable shale barrier resides and extends between the two different fluid-bearing reservoir strata throughout the entire oil reservoir.

If crossflow does exist between the strata of differing permeability, then this type of matrix-rock vertical conformance problem proves to be a difficult conformance problem to successfully remedy (or partially remedy). Also, this type of conformance problem is often a costly one to successfully remedy in full or in part. When treating this conformance problem, conformance-improvement operations must be extended and applied deep (far-wellbore) into the reservoir—often over half of the distance between an injection and production well. This conformance problem is usually treated with a conformance-improvement operation involving the application of an enhanced-viscosity drive-fluid flooding operation (e.g., polymer flooding).

As illustrated in **Fig. 4.4,** when crossflow occurs between reservoir strata of differing permeability, the selective placement of a permeability-reducing agent or a blocking agent near the wellbore in the high permeability strata of an injection or production well (even if that could be achieved—see Section 4.2.5) will provide only temporary conformance improvement. In the case of an injection-well treatment (Fig. 4.4a), after the oil-recovery drive fluid radially penetrates in the low permeability strata, it will rapidly crossflow from the low-permeability strata to the high-permeability strata once it reaches the outer penetration distance of the permeability-reducing material placed in the high permeability strata. As a result, little conformance improvement will be

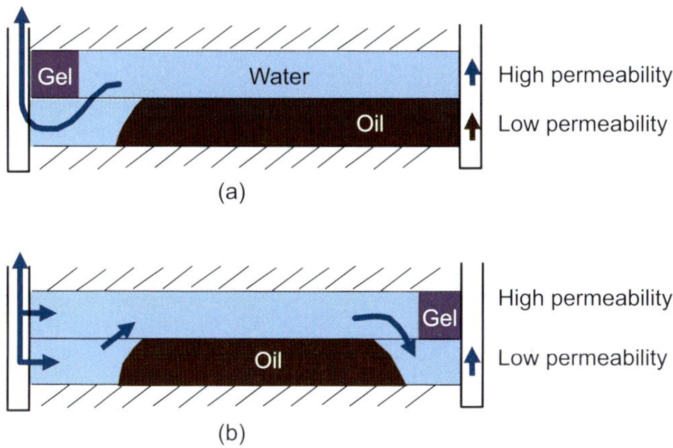

Fig. 4.4—For matrix-rock vertical conformance problems with crossflow, near-wellbore permeability-reducing treatments are not highly effective.

achieved and insignificant incremental oil production will be realized from such a conformance-improvement treatment.

As also can be seen in Fig. 4.4b, relatively low benefit and incremental oil production will be realized from a permeability-reducing or blocking-agent water-shutoff treatment selectively placed near-wellbore in the high-permeability strata at a production well. This is because just outside of the outer radial penetration of the placed water-shutoff treatment material, the water flowing in the high-permeability strata crossflows from the high-permeability strata into the low-permeability strata. In such a radial flow situation, only a small volume of oil residing in the low-permeability strata will be recovered by the water-shutoff treatment.

Water-shutoff treatments that have been applied to reservoirs where the conformance problem involves strata of differing permeability overlaying one another with crossflow between the geological strata have caused a number of oilfield operators to misdiagnose why such water-shutoff treatments did not last as long as expected and did not generate as much incremental oil as was expected and desired. The operators, in this case, historically have tended to conclude that the water-shutoff blocking-agent material broke down chemically or physically very quickly with time, when in fact, the water-shutoff blocking-agent material did not do so. The rapid crossflow of water from the high-permeability strata to the low-permeability oil-bearing strata (Fig. 4.4b) caused the relatively fast re-establishment of the high-watercut production. This misdiagnosis of the performance of such water-shutoff treatments often results because the operator does not realize that crossflow is occurring between the reservoir strata of differing permeability or does not understand how such crossflow will limit the effectiveness and performance of water-shutoff treatments.

As previously noted, if the operator of an oil field is not sure whether the impermeable geological stratum separating the oil-bearing reservoir strata of differing permeability is, or is not, continuous within the reservoir or within a well pattern, then it is normally recommended that the operator assume (erring on the conservative side) that the oil-bearing strata of differing permeability are in fluid and pressure communication.

There are techniques such as well testing available to determine whether reservoir strata of differing permeability are in fluid-flow and pressure communication with one another (see Table 4.1).

4.8 Near-Wellbore vs. Far-Wellbore Conformance Remedies

Conformance-improvement operations can be conducted in the near-wellbore region (including within the wellbore itself) and in the far-wellbore reservoir region. The placement and penetration required of the conformance-improvement treatment depends on the target conformance problem (Seright et al. 2003; Sydansk 2007).

Conformance-improvement techniques and operations performed in the near-wellbore region are, in general, less expensive, easier, and less risky to apply compared to conformance-improvement treatments performed in the far-wellbore region. Examples of conformance problems that can be successfully applied in the near-wellbore environment are:

- Reservoir strata without crossflow (Section 4.2.4)
- Water channeling behind pipe (Section 4.2.9)
- Casing leaks (Section 4.2.10)

Examples of conformance problems that must be treated in the far-wellbore environment are:

- Viscous fingering (Section 4.2.1)
- Directional areal permeability trends in matrix-rock reservoirs (Section 4.2.2)
- Reservoir strata with crossflow (Section 4.2.3)
- Water coning up through fractures (Section 4.2.6)
- Fracture channeling (Section 4.2.7)
- Solution channels and interconnected vugular porosity (Section 4.2.8)

The conformance problems of water coning up through fractures (Section 4.2.6) and fracture channeling (Section 4.2.7) involve fracture conformance problems. Such conformance problems can be successfully treated with intermediate-sized volumes (typically 100 to several 10,000 bbl) of an inexpensive permeability-blocking agent (e.g., a polymer gel) that is placed only in the limited and small volume of the open fractures that are treated (as compared to the total matrix-rock pore volume).

4.9 Underestimation of the Permeability of High-Permeability Flow Paths

A serious problem in the successful application of many conformance-improvement treatments and operations for high-permeability flow channels is the underestimation of the permeability of the flow channels that exist within the reservoirs (Sydansk 2007; Sydansk and Southwell 2000).

Many conformance-improvement materials work within a limited permeability range. If the incorrect type of conformance-improvement-treatment material is chosen in this respect, then the conformance-improvement treatment is often doomed to failure.

The recommended design strategy and treatment selection for an oilfield operator when applying a conformance-improvement operation in a reservoir where no conformance-improvement operations have been previously applied, is to use the highest estimate of the permeability of the flow channels.

Chapter 5

Excessive Water Production

5.1 Excessive Water Production as a Conformance Problem

Excessive water production due to conformance problems becomes an issue when it competes directly with oil production. This water usually flows to the wellbore through its own path, independent of the oil flow pathway. In such cases, a reduction of water production can often lead to a greater pressure drawdown and increase oil production rates (Seright et al. 2003).

The purpose of this chapter is to introduce the different conditions for excessive water production during oil-recovery operations and to establish when excessive-water production becomes a conformance problem and a major industry challenge. Therefore, different types of excessive-water production are presented and briefly discussed.

5.1.1 Excessive Water Production as a Conformance Problem: A Hypothetical Case.
For the sake of illustration, it will be assumed that a waterflooding operation is being conducted in a 5-spot well pattern where the production well with a casing-leak excessive-water-production problem is located. It is further assumed that waterflood water breakthrough has not occurred yet at any of the production wells of the 5-spot well pattern.

As previously noted in this book, conformance is defined as a measure of the degree of sweep efficiency during an oil-recovery flooding operation. Thus, it would probably appear, at first glance, that for the on-going process of oil-recovery flooding, a casing-leak excessive-water-production problem occurring at one of the production wells does not negatively impact the sweep efficiency of the oil-recovery flooding operation. However, closer consideration shows that indeed it can; and, the negative impact can create a conformance problem.

As a result of the casing leak being so extensive (because the water-bearing geological strata is allowed to flow at a nearly infinite rate of water through the casing leak), the well produces at nearly 100% water cut—as compared, in this hypothetical case, to producing at nearly 100% oil cut if there were no casing leak. Therefore, the production well with the casing-leak excessive-water-production problem does not produce "any" oil and the sweep efficiency of the oil-recovery process is negatively impacted because the waterflood front towards the production well that has the casing leak is substantially impaired and retarded. Excessive water production during oil-recovery operations directly competes with oil production. Thus, each barrel of water that is produced replaces a barrel of oil that is not produced.

5.1.2 Conditions of Excessive-Water-Production Conformance Problems. The preceding example is extreme; the aim is to show the reader that a casing-leak excessive-water-production problem occurring in a production well of any well pattern undergoing an oil-recovery flooding operation (any secondary or tertiary oil recovery operation) can negatively affect the sweep efficiency of the process—and thus create at least some degree of a conformance problem.

Excessive-water-production problems can also occur during primary oil production and negatively impact oil-recovery efficiency. For example, it is not uncommon for a well to suffer from excessive and unnecessary water production resulting from water that is coning up through open vertical fractures to the well from an underlying aquifer.

Excessive-water-production conformance problems resulting from casing leaks, water coning through matrix rock (see Section 4.2.5), and water channeling behind pipe (see Section 4.2.9) are not necessarily caused by reservoir permeability heterogeneity.

In contrast, other excessive water production conformance problems that occur during oil-recovery operations can be, and often are, directly attributable to reservoir permeability heterogeneity. For instance, consider the excessive-water-production conformance problem involving direct fracture channeling between an injection well and a production well in a 5-spot pattern during a waterflooding operation—as illustrated in Fig. 4.1g. Assume that the fracture fluid-flow conductivity and permeability are so great that, in turn, the affected production well is producing such a high water cut and high water/oil ratio that the well has to be shut in as a result of having become uneconomic. This would be a clear-cut case of an excessive-water-production problem that is directly caused by reservoir permeability heterogeneity.

Examples of types of conformance problems that can cause excessive water production due to reservoir permeability heterogeneity, negatively impacting oil-recovery-flooding sweep efficiency include:

- High-permeability matrix-rock directional trends (see Section 4.2.2)
- High-permeability matrix-rock strata with crossflow (see Section 4.2.3)
- High-permeability matrix-rock strata without crossflow (see Section 4.2.4)
- Fracture channeling (see Section 4.2.7)
- Solution channels and interconnected vugular porosity (see Section 4.2.8)

Excessive water production during oil-recovery operations adds unnecessary operating-expense costs to the recovery of oil. The unnecessary operating expenses include water lifting, treating, handling, and disposal costs; they can also include environmentally-related and safety-related costs derived from, for example, possible heavy-metals and H_2S being produced in the excessive and unnecessary water production.

5.2 Necessary and Unnecessary Water Production

There are two distinct classes of oilfield water production as related to conformance and conformance problems; they are *necessary* and *unnecessary* water production (Sydansk 2007).

5.2.1 Necessary Water Production. Necessary water production is also referred to as *good water production*. Necessary water production involves fractional flow of water and oil through a matrix-rock reservoir to the production well. During a waterflood in a "homogeneous" matrix-rock reservoir, water displaces and mobilizes oil through fractional oil/water flow. After water breakthrough occurs at the production well, simultaneous oil/water production takes place.

For every barrel of necessary water production that is shut off in this situation, the oil production rate will consequentially also be reduced by some amount. The previous statement applies to a given geological structure or stratum that is producing at a given fractional flow. In many cases, oil-producing operators are very hesitant about giving up any substantial oil-production rate—especially as it relates to conducting a water-shutoff treatment. Thus, reducing the production rate of necessary water is usually not considered an appropriate approach for the application of conformance-improvement water-shutoff treatments.

Conformance problems are not the cause of, and do not promote, necessary water production. This is true even when nearing the end of the economic life of a waterflood that is being conducted in a homogeneous reservoir. At this point, the water/oil ratio (WOR) and water cut are becoming excessively high. This serves as a signal that the waterflood and the reservoir have reached the point at which all moveable oil in the reservoir or well pattern that can be recovered economically has been recovered. In this situation in which undesirably large amounts of water are being produced, the reservoir or well pattern has been fully swept by the waterflood and the economic limit of the waterflood has been reached.

5.2.2 Unnecessary Water Production. Unnecessary water production is also referred to as *bad water production*. Unnecessary water production usually involves the *flow of water and oil via separate flow paths to the producing interval of the wellbore.*

An example of unnecessary water production is water coning up from an underlying aquifer. In this case, the water flows to the wellbore by its own flow path from the aquifer, and the oil flows through its own separate path from the oil-producing reservoir strata to the wellbore. Another example of unnecessary water production is direct channeling of water from an injection well to the production well through highly conductive natural fractures that run directly from the injection well to the production well. In this case, during waterflooding, the water flows (for the most part) directly from the injection well to the production well through the separate fracture flow path. In this situation, the oil is produced (for the most part) directly or indirectly into the production well via a separate reservoir flow path—namely, flowing mostly through the reservoir matrix rock.

Normally in these cases, shutting off or reducing the rate of unnecessary water production is the prime focus of oilfield conformance-improvement water-shutoff treatments. Conformance problems are the root cause of unnecessary water production.

5.3 Strategy to Successfully Apply a Water-Shutoff Treatment

When considering the application of a conformance-improvement water-shutoff treatment, it is strongly recommended that the operator start by identifying (or deducing as best as the operator can) the source and flow path to the wellbore of the water that needs to be shut off (e.g., water in an underlying aquifer).

TABLE 5.1—EXCESS-WATER-PRODUCTION PROBLEM CATEGORIES AND CORRESPONDING TREATMENT OPTIONS IN INCREASING ORDER OF PROBLEM TREATMENT DIFFICULTY (Seright et al. 2003)

Problem Category	Problem Types	Treatment Options
Category A	1. Casing leaks without flow restrictions 2. Flow behind pipe without flow restrictions 3. Unfractured wells (injectors or producers) with effective barriers to crossflow	1. Cement or mechanical patches 2. Cement or a combination of gel placement followed by cement to fill and plug larger near-wellbore voids and to prevent gels from washing out from their placement 3. Cement or sand plugs are commonly used to stop water production when the water zone is located at the bottom of the well. If the water zone is located above an oil zone, the most common water-shutoff methods include cement or carbonate squeezes (into perforations) or mechanical packers or patches
Category B	4. Casing leaks with flow restrictions 5. Flow behind pipe with flow restrictions 6. "2D coning" through a hydraulic fracture from an aquifer 7. Natural-fracture system leading to an aquifer	4. Gels that can flow easily through the small casing leaks and some distance into the formation during gel-treatment placement, but the gel must be rigid after setting 5. Gelants that can flow or extrude readily through narrow constrictions. It is important that the gelant penetrate into the permeable matrix rock adjacent to the channel behind the pipe 6. Gelant treatments that flow along the fracture and leak off a short, predictable distance into the matrix of all zones (water, oil, and gas); the gel should reduce the permeability to water much more than the permeability to hydrocarbon in the treated matrix rock 7. Treatments using partially formed gels or preformed gels
Category C	8. Faults or fractures crossing a deviated or horizontal well 9. Single fracture causing channeling between wells 10. Natural-fracture system allowing channeling between wells	8. Selective placement of preformed gel 9. Conventional gel treatments and preformed gels; the volume of gel to be injected depends critically on fracture width and gel properties (i.e., gel composition and rigidity); in cases in which wide fractures or large vugs are present, the mechanical-strength and plugging characteristics of the gel can be increased by adding particulate matter (e.g., sand, cellophane, fibers, nut shells) 10. Preformed gel treatments to reduce injector-producer channeling in naturally fractured reservoirs can be applied in either injection or production wells
Category D	11. 3D coning 12. Cusping 13. Channeling through strata (no fractures), with crossflow	Gelants and gel treatments are ineffective for treating Problems 11, 12, and 13, which are the most difficult conformance problems listed here. A practical fix to the water coning problem is to decrease the fluid-production rate, which might not be cost effective in the long term. The only long-term, practical solution to stop water coning production is to plug the zone. If gelant is used to plug the offending zone, the hydrocarbon productive zones must be protected during gelant placement (Seright et al. 2008).

As previously indicated, in the case of excessive water production due to multiple scenarios, a strategy recommended by Seright et al. (2003) is to treat first the easiest excessive-water-production problems. A prioritization and categorization of various types of water problems from least to most difficult with their corresponding treatment options, prepared by Seright et al. (2003), is presented in **Table 5.1.**

During the decision making process, the operator needs to keep in mind that for any given conformance-excessive-water-production problem type, more than one of the numerous water-shutoff treatments and technologies that are available may shut off the excessive water production (Table 5.1). The operator should address the easiest problems first and carefully design (and properly size) the water-shutoff treatment or operation that will maximize the probability of achieving economic success for the treatment.

5.4 Excessive Water Production Is a Major Industry Challenge

It has been suggested that the oil industry should begin to ask itself: *Are we oil-producing or water-producing companies?*

According to a 2004 report from a U.S. national laboratory (Veil et al. 2004), the oil industry was, on the average, producing worldwide *3 bbl of water for every 1 bbl of oil.* This translates to an average worldwide water cut production of 75%.

The same report states that the oil industry in the United States was producing *7 bbl of water for every 1 bbl of oil,* which is equivalent to an average of 88% water cut. Many oil fields in the Rocky Mountain region and in the Permian Basin of Texas are producing, on the average, at water cuts exceeding 95%, and some at water cuts exceeding 98%. This is a tremendous amount of water produced for every barrel of oil produced—especially water that is costly to produce and dispose of, and water that can be an environmental liability. The operating expenses of oilfield water production include costs of water lifting, handling, treatment, and disposal. Oilfield produced water can be an environmental liability for a number of reasons, including the presence of dissolved H_2S and toxic heavy metals in the produced water. **Table 5.2** summarizes average industry-wide WOR and water-cut percentages for oil production on a worldwide basis and in the United States.

As noted in the cited 2004 government report (Veil et al. 2004) and noted elsewhere, the future trend for the oil industry will be to produce, on the average, at even substantially higher water cuts.

A positive note here is that there is a possibility that the industry might find and develop some economically viable and socially beneficial uses, in the not too distant future, for some of the water that is being produced. There have been tremendous strides made over the past several decades in the development of micro- and nano-filtration technologies. It is possible that in the somewhat near future, technologies will

TABLE 5.2—OVERVIEW OF THE OIL-INDUSTRY AVERAGE WATER PRODUCTION (Veil et al. 2004)

	bbl water/bbl oil	Water cut
Worldwide	3:1	75%
USA	7:1	88%

become available to economically clean up produced water to the point that it can be reliably and safely used for industrial and agricultural uses—and even, conceivably, for potable water use.

It is considered that much of (and probably the majority of) the tremendous amount of water that the oil industry is presently (2011) coproducing with oil results from conformance problems. Thus, there is not only a great potential for conformance-improvement operations to render large amounts of improved and incremental oil production and associated increased revenues for the oil industry, but conformance-improvement operations also can be used to save the industry large amounts of oil-production operating expenses through the application of effective conformance-improvement water-shutoff treatments and operations that are described in the subsequent chapters of this book.

Chapter 6

Improving Conformance—A Broad Overview

This chapter briefly reviews means, methods, techniques, operations, and treatments that are used to improve conformance during oil-recovery operations (Seright et al. 2003; Sydansk 2007).

6.1 Types of Conformance-Improvement Methods

Table 6.1 presents a list of the various conformance-improvement methods. This list should not be considered an exhaustive enumeration of all of the various types of conformance-improvement methods.

6.2 General Means to Improve Conformance

In this section, the general means that are employed to improve conformance in conventional oil reservoirs will be enumerated and briefly discussed.

6.2.1 Increasing the Viscosity of the Flooding Fluid. Increasing the viscosity of the oil-recovery drive fluid during an oil-recovery flooding operation (e.g., polymer flooding) is an effective means to promote conformance improvement within matrix-rock oil reservoirs. This is presently (2011) the most widely used means to improve conformance deep (far wellbore) within the matrix rock of a reservoir, as is required when treating, for example, the conformance problems of viscous fingering (see Section 4.2.1) or reservoir geological strata of differing permeability with crossflow (see Section 4.2.3). The viscosity-enhancement conformance-improvement technologies involve large-scale and large-volume flooding operations. Improving conformance by increasing the viscosity of the oil-recovery drive fluid during an oil-recovery flooding operation is discussed in more detail in Chapter 7.

6.2.2 Reducing the Permeability of High-Permeability Flow Channels. Reducing the permeability or totally plugging reservoir high-permeability flow channels and/or anomalies is the second most widely applied method to promote conformance improvement within the oil reservoir itself. This means of improving conformance is applied within the rock of the reservoir and is not applied in or at the wellbore.

However, this method can at times be applied in reservoir rock of the near-wellbore region. The permeability-reduction scheme for improving conformance normally involves application of conformance-improvement treatments that are of relatively small volume as compared to a conformance-improvement viscosity-enhancing flooding operation (e.g., polymer flooding). Improving conformance through reduction of the permeability of the reservoir high-permeability flow channels is discussed in more detail in Chapter 8.

6.2.3 Increasing the Permeability of Low-Permeability Flow Channels. Another method to improve conformance within the reservoir rock phase is to increase the permeability and/or the fluid-flow capacity of reservoir low-permeability flow paths. For conformance-improvement purposes the application of this technique is mostly limited to the relatively near-wellbore region of injection wells.

Conformance-improvement means that involve increasing the permeability of low-permeability reservoir flow paths are beyond the scope of this book. The strategy of increasing the permeability of low-permeability oil-reservoir flow paths is probably

TABLE 6.1—REPRESENTATIVE TYPES OF CONFORMANCE-IMPROVEMENT METHODS	
Method	**Further Information**
• **Cement (Portland)**	**Section 6.2.4**
➢ Squeeze cementing	Section 6.2.4
➢ Foamed cement	Section 6.2.4
➢ Microfine cement	Sections 6.2.4 and 8.2.4
➢ Grey-water cement solutions	Section 6.2.4
➢ Cement containing specialty chemicals	Sections 6.2.4 and 8.2.4
• **Flooding with viscous fluids**	**Section 6.2.1; Chapter 7**
➢ Polymer flooding	Chapter 7
• **Permeability-reducing treatments**	**Section 6.3/6.2.2; Section 7.3; Chapter 8**
➢ Gels	Section 8.2.1
▪ Inorganic-based bulk gels	Section 8.2.1
○ Silicate gels	Section 8.2.1
▪ Organic-based polymer bulk gels	Section 8.2.1
○ Polymers	Section 7.2.2
◆ Synthetic	Section 7.2.2
◆ Biopolymers	Section 7.2.2
○ Crosslinking agents	Section 8.2.1
◆ Inorganic	Section 8.2.1
◆ Organic	Section 8.2.1
▪ Organic-monomer-based in-situ-polymerized gels	Section 8.2.1
▪ Preformed polymer-gel particles	Sections 8.2.1/8.2.4
○ Microgel particles	Sections 8.2.1/8.2.4
○ Delayed-swelling ("popping") microgel particles	Section 8.2.1
○ Colloidal dispersion gels	Section 8.2.1
○ Preformed swelling gel particles (for treating high-permeability anomalies)	Section 8.2.1
➢ Resins	Section 8.2.3
➢ Specialty polymers alone [for relative permeability modification (RPM) water-shutoff treatments]	Section 8.2.2

TABLE 6.1—REPRESENTATIVE TYPES OF CONFORMANCE-IMPROVEMENT METHODS (Cont.)	
Method	**Further Information**
• **Foams and foam flooding** (special case—both viscosity enhancing *and* permeability reducing)	**Section 7.6 ; Chapter 9**
➢ Conventional foams	Sections 9.1 to 9.8
➢ Polymer-enhanced foams	Section 9.9
➢ Foamed gels	Section 9.9
• **Particulates**	**Section 8.2.4**
• **Mechanical wellbore methods**	**Section 6.2.4**
➢ Packers and bridge plugs	Section 6.2.4
➢ Straddle packers	Section 6.2.4
➢ Sliding sleeves	Section 6.2.4
➢ Tubing patches	Section 6.2.4
➢ Sand-back plugs	Section 6.2.4
• **Wellbore drilling and completion methods**	**Section 6.2.5**
➢ Selective completion and selective perforating	Section 6.2.5
➢ Use of horizontal and multilateral wellbores	Section 6.2.5
➢ Use of intelligent wells and well completions	Section 6.2.5
➢ Use of wells that can be selectively "snaked" through the reservoir	Section 6.2.5
• **Well locating**	**Chapter 6, Section 6.2.5**
➢ Strategic and optimum areal placement of vertical wells	Section 6.2.5
➢ Strategic well pattern selection and placement	Section 6.2.5
➢ Strategic and optimum placement vertically and directionally of horizontal wells	Section 6.2.5
• **Infill drilling**	**Section 6.2.5**
• **Well abandonment and selective shut-ins**	**Section 6.2.5**
• **Pattern balancing, well realignment, and shut-ins**	**Section 6.2.5**
• **Comprehensive reservoir description**	**Section 6.2.5**
• **Increasing the permeability** (normally relatively near wellbore) of low-permeability flow paths	**Section 6.23**
➢ Acidizing	Section 6.2.3
➢ Selective hydraulic fracturing	Section 6.2.3
➢ Deep perforating	Section 6.2.3

better suited for discussion in the context of production-enhancing stimulation technologies. Application of permeability-stimulation techniques to low-permeability reservoir strata for the *sole purpose* of improving conformance is not widely applied.

Using any one of the three following techniques for selectively stimulating (increasing) the permeability of low permeability strata in oil reservoirs is a potential means of improving conformance:

- Matrix-rock acid stimulation
- Selective (and usually small scale) hydraulic fracturing
- Deep perforation

In order for these three conformance-stimulation techniques to be effective, there should be no fluid-flow and pressure communication between the reservoir geological strata of differing permeabilities. In addition, the well to be treated should be injecting oil-recovery drive fluid (e.g., water) in the radial-flow mode.

6.2.4 Improving Conformance Through the Wellbore. There are numerous means and techniques that can be highly effective for use in improving conformance in conventional oil reservoirs, where the methods and techniques are not applied within the rock of an oil reservoir. Although these techniques and technologies can be very powerful and effective to improve conformance, they will not be discussed further in this book beyond what is discussed in this section.

Mechanical Methods. Mechanical methods that are applied at, or within, the wellbore can be used to treat certain conformance problems—especially matrix geological strata of differing permeability overlying one another when there is no crossflow between the geological strata.

The use of mechanical means for improving conformance, when there is a good match between the conformance problem and the mechanical treating method, is favored for the following reasons. Mechanical methods are often relatively inexpensive and have a relatively low risk factor. They also are often relatively simple and straightforward to apply. Wellbore mechanical methods used to improve conformance include the use of packers and bridge plugs, straddle packers, sliding sleeves, tubing patches, and sand-back plugs.

Cement. Cementing technologies that can be employed for conformance-improvement purposes include squeeze cementing, foamed cement, microfine cement, cement containing specialty chemicals or materials, and "gray water" cement. Cement can be applied within the wellbore, or in the extended wellbore (e.g., perforations), as a material and as a mechanical method to treat conformance problems resulting from matrix-rock geological strata of differing permeability overlying one another when there is no crossflow and pressure communication between the geological strata. Portland cement placement in the wellbore also has been successfully used to promote *temporary and short-term* water-shutoff when water-coning production first begins.

Portland Cement. Portland cement is widely promoted and used as a means to improve oil-reservoir conformance problems. The use of Portland cement as a conformance-improvement technique could have possibly been included in the mechanical-methods discussion above. But, in view of the unique features of improving conformance through the application of cement and the fact that the use of cement for improving conformance is one of the oldest and most widely applied conformance-improvement techniques, the authors have chosen to discuss separately the use of cement for improving conformance.

Portland cement for conformance-improvement application in vertical wells is essentially limited to its use as a fluid-flow-blocking agent when the cement material is placed either in perforation tunnels or, alternatively, in a vertical wellbore to shut off a lower-producing interval at, or near, the bottom of the well.

Portland cementing for squeeze cementing can be used for conformance-improvement purposes when the cement is placed in:

(a) perforation tunnels when treating matrix strata problems without crossflow
(b) water channels behind pipe when large flow-channels occur [≥4-mm (>0.16-in.) openings]

   (c) casing leaks having large casing-leak flow channels [≥1–2-mm (>0.04–0.08-in.) openings]

To repair "tight" casing leaks and small gaps or fractures in cement, microfine cements are employed. Microfine cements are of low particle sizes, having an average particle size of 6 µm and a maximum particle size of 10 to 15 µm. The fine particles have a greater penetration than conventional oil well cements and can thus enter the leak (Clarke et al. 1993; Boisnault et al. 1999; Bensted 2002).

A common misconception that oilfield operators have is that cement material can be placed relatively deeply, in certain instances, into reservoir matrix rock or into reservoir high-permeability anomalies, such as fractures or solution channels. However, Portland cement materials are not, and cannot, be injected below formation parting pressure into, and placed any significant distance (more than an inch or so at the most) into matrix-rock reservoir of normal permeabilities (<10 darcies). This includes microfine cement. In addition, conventional Portland cement cannot be placed deeply (often <10 ft) into reservoir high-permeability anomalies (e.g., fractures or solution channels) having reservoir flow apertures/widths of <4 mm (<0.16 in.). Portland cement can only be injected into, and placed within, matrix-rock reservoir if its injection pressure exceeds the breakdown (parting) pressure of the matrix rock. There are few, if any, cases in which the injection of Portland cement into matrix reservoir rock at injection pressures exceeding reservoir parting pressure can be used to promote the successful treatment of a reservoir conformance problem.

During Portland cement applications, "non-damaging and non-permeability-reducing" cement filtrate loss from Portland cement can leak off under high differential-pressure conditions into adjacent reservoir matrix-rock if the cement filtrate does not contain specially added gelling chemicals.

"Gray Water" Cement. Gray water cement involves the injection of an extremely dilute concentration of Portland-cement particles in which the injected solution appears to be only a grayish-colored and cloudy, watery solution. In terms of conformance improvement, gray water cement is used to attempt a deeper penetration (more than a few inches or so at the most) into the matrix-rock reservoir and it might be better categorized as the application of a solid-particle conformance-improvement technology.

Over-Use of Cementing. Partly as a force of habit and possibly with the encouragement of some oilfield service companies, the first technology that oilfield operators often consider using when encountering any conformance problem is the application of cement. The successful application of cement technologies for conformance-improvement purposes is not a panacea. In fact, cement technologies can only be successfully applied to remedy a limited number of the various oil-reservoir conformance problems. Because of this and other limitations discussed in this section, the track record for the successful application of cement technologies to remedy oil-reservoir conformance problems has not been exceptionally good.

6.2.5 Other Methods for Circumventing Conformance Problems. *Infill Drilling.*
Infill drilling, which is the drilling of wells in-between the existing wells, is another powerful method that can be used to improve conformance within oil reservoirs and to generate improved oil recovery (IOR), even though it can be a relatively costly means.

Infill drilling is widely applied as a method to improve conformance because infill wells are usually drilled at various spacings and coupled with waterflooding and enhanced-oil-recovery (EOR) processes to increase areal sweep and to reduce the detrimental effects of rock heterogeneities that may lead to bypassing of the oil (Satter et al. 2008). On the positive side, this is a relatively low-risk conformance-improvement technique. Normally, the most prevalent questions that revolve around infill drilling for conformance improvement are economic in nature and focus on above-ground environmental issues. The fundamental limitation, relating to how closely wells can be spaced in order to improve conformance, is most often considered to be economic in nature.

Pattern Balancing, Well Realignment, and Shut-Ins. During any flooding operation (e.g., waterflooding and/or polymer flooding), the areal pattern consisting of a number of wells—typically five to nine—is balanced by manipulating injection and production rates. In a perfectly balanced pattern, the driving fluid (e.g., injected water) does not migrate outside the pattern, thereby ensuring maximum areal-sweep efficiency and—along with well realignment—can result in higher oil recovery from the pattern area (Satter et al. 2008). Therefore, pattern balancing and realignment and selective well shut-ins are widely applied means of promoting conformance improvement.

Comprehensive Reservoir Description. Comprehensive geological and petrophysical reservoir description together with continuous reservoir monitoring for the explicit purpose of improving conformance are both techniques that can be used to generate significant volumes of improved and incremental oil recovery from of oil reservoirs.

Well and Well-Pattern Locating. Strategic and optimized well and well-pattern locating is a conformance-improvement technique that can prove to be highly effective in certain instances. For an oil field involving vertical wells, this technique can be effectively used to reduce conformance problems when encountering directional areal permeability trends in matrix-rock oil reservoirs. This is another conformance-improvement technique that is widely applied. As in the case of infill drilling, the critical issue relating to well and well-pattern locating in order to improve conformance is considered to be economic in nature.

Horizontal Wells and Advanced Well Technologies. Horizontal wells and other advanced well technologies are procedures that are *rapidly evolving* to become very effective and widely applied techniques to minimize or circumvent many oil-reservoir conformance problems, as are the use of advanced high-tech and smart well completions.

In conjunction with the use of advanced reservoir description techniques that are conducted prior to drilling the well, the use of advanced drilling techniques in which the well can be accurately "snaked" virtually at will throughout any given oil reservoir is an emerging conformance-improvement technique to circumvent many oil-reservoir conformance problems (Talukdar and Instefjord 2008; Blount et al. 2006; Smith and Perepelecta 2002). However, this conformance-improvement technique is not readily (and not likely) applicable to existing and producing oil reservoirs where the reservoir has already been developed on relatively tight well spacing.

6.3 Definition of Conformance-Improvement Treatment

As has been historically used by petroleum engineers practicing conformance-improvement technologies, the term "conformance-improvement treatment" has a somewhat restricted connotation and definition. The term conformance-improvement treatment tends to be reserved for the application of relatively small-volume (1 to 50,000 bbl) *permeability-reducing* conformance-improvement agents—for instance polymer-gel or resin treatments.

Although a large-volume polymer-flooding *viscosity-enhancing* oil-recovery operation is, in a sense, treating an oil-reservoir conformance problem, it is considered a flooding operation and not a "conformance-improvement treatment." Again, the term "conformance-improvement treatment," *in practice,* has been limited to the application of relatively small, permeability-reducing conformance-improving treatments.

An exception to the above-stated volume range for permeability-reducing gel treatments are several microgel injection-well sweep-improvement technologies that can involve the injection of several 100,000 bbl of treating fluid. In this case, it is not clear whether these microgel sweep-improvement technologies should be considered as the application of gel treatments or as flooding operations (see Sections 8.1 and 8.2.1 for more details on this topic).

6.4 Matching Conformance Problems to Conformance-Improving Technologies

Critical to successfully remedying, or partially remedying, a given oil-reservoir conformance problem is matching the conformance problem with a conformance-improvement operation or technique that can be successfully used to remedy the reservoir conformance problem.

The critical factor here is that a given oil-reservoir conformance problem can often only be successfully mitigated by a limited number of techniques chosen from the large number of conformance-improvement technologies and techniques that are available. Thus, if the conformance problem is not correctly diagnosed or if an inappropriate and ineffective conformance-improvement technology or technique is chosen to be applied to the conformance problem, then the conformance-improvement operation is almost certainly doomed to failure.

6.5 Benefits of Conformance Improvement

The benefits of conformance improvement (whether conducted in the reservoir, the wellbore, or elsewhere) in conventional oil reservoirs can be exceptionally large and lucrative in terms of improved and incremental oil recovery and in terms of incremental oil-production revenues to the operator. The benefits of conformance improvement in oil reservoirs can include generating:

- Better sweep efficiency during oil-recovery flooding operations
- Increased and incremental oil recovery (possibly recovered in a highly profitable manner)
- IOR (possibly recovered in a highly profitable manner)
- Accelerated oil-recovery rates and improved business profitability (time value of money)
- Reduced oil-production expenses (OPEX) through the reduction of unnecessary recycling of driving fluids (e.g., water and/or polymer solution) and the associated lifting, handling, treatment, environmental-related, and disposal costs

- Extended economic lives of certain oil reservoirs that are approaching their economic limit
- Reduced environmental liabilities and increased environmental benefits (e.g., produce less saline, heavy-metal-containing, and/or H_2S-containing produced water)

In addition, conformance-improvement treatments and technologies (e.g., polymer gels) can be used to successfully treat and reduce the fluid-flow capacity of anomalous highly conductive fractures in highly naturally fractured carbonate reservoirs, where imbibition and gravity drainage are the primary oil-recovery mechanism. In this case, the conformance-improvement treatments permit the oil-recovery flooding fluid (e.g., water) to more effectively sweep and recover, from all the reservoir fractures, the oil that has been produced from the individual matrix-rock reservoir blocks into the adjacent fractures.

Chapter 7

Improving Conformance by Increasing Viscosity

7.1 Increasing the Viscosity of a Flooding Fluid to Improve Conformance

As briefly noted previously, the following conformance problems can only be successfully treated with conformance-improvement techniques that function within the oil reservoir at some distance from the wellbore:

- Mobility-control viscous fingering
- Directional areal high-permeability trends in matrix rock
- Matrix strata of differing permeability overlying one another in which crossflow can occur
- Fracture channeling
- Coning through vertical fractures
- Solution and interconnected-vugular channels (see Section 4.2.8)

There are two distinct conformance-improvement techniques that can be applied within the oil-reservoir rock itself to successfully treat these conformance problems (Sydansk 2007). The first technique is to increase the viscosity of the oil-recovery drive fluid during oil-recovery flooding operations that are being conducted in matrix-rock reservoirs. The second technique is the placement of a permeability-reducing material in the offending reservoir high-permeability flow channels.

This chapter addresses the technique of increasing the viscosity of the oil-recovery drive fluid during oil-recovery flooding operations that are conducted in matrix-rock reservoirs.

Increasing the viscosity of the oil-recovery flooding fluid (e.g., water) as a conformance-improvement technique improves the sweep efficiency by two different mechanisms. First, the increased viscosity of the oil-recovery drive fluid can be exploited to *improve the mobility ratio* during an oil-recovery flooding operation (see Section 2.1). This viscosity-enhancing mechanism reduces the viscous-fingering conformance problem for the oil-recovery flooding operation. Second, the increased viscosity in the oil-recovery drive fluid can improve sweep efficiency and conformance *when flooding heterogeneous matrix reservoir rock* (permeability heterogeneity). The increased viscosity of the oil-recovery drive fluid used in a conformance-improvement technique does not normally increase the flood's capillary number such that the flood's microscopic displacement efficiency is increased to the point that residual oil saturation is recovered.

This conformance-improvement technique functions not only in the near- and intermediate-wellbore region of a reservoir, but also in the far-wellbore region. As such, this technique normally involves the injection of relatively large volumes of the conformance-improvement fluids as *oilfield flooding operations* that can go on for long periods of time (e.g., possibly a number of years) and are often applied field-wide or to a large portion of a field. This conformance-improvement technique usually requires injecting volumes of the viscosity-enhanced oil-recovery drive fluid that are measured in terms of *pore volumes* of the reservoir volume being treated. At the present time (2011), a one-half pore volume or more is not uncommon for viscosity-enhancing polymer-flooding conformance improvement.

A viscosity-enhanced flooding operation often involves the injection of the conformance-improvement fluid simultaneously into a number of adjacent injection wells; it is always applied exclusively from the injection-well side.

7.2 Polymer Flooding To Increase Viscosity

7.2.1 Introduction to Polymer Flooding. Any means that can be used to increase the viscosity of the oil-recovery drive fluid (e.g., water) during an oil-recovery flooding in reservoir matrix rock will help to improve the sweep efficiency and the conformance of the flooding operation.

Polymer flooding is the most widely applied conformance-improvement technology for increasing the viscosity of the drive fluid during oil-recovery flooding in a matrix-rock oil reservoir (Lake 1989; Sorbie 1991; Green and Willhite 1998; Sydansk 2007; Wang et al. 2008b). During polymer flooding, low concentrations (e.g., 500 to 2,000 mg/L) of a water-soluble and high-molecular-weight polymer are dissolved in the injection water to increase its viscosity. Polymers are large molecules and chemical entities that are referred to as macromolecules. Polymer molecules are the resultant chemical specie when a large number of relatively small and repeating molecular entities, called monomers, are chemically joined together. The chemical process of joining together the monomers and forming polymer molecules is referred to as the polymerization process.

7.2.2 Biopolymers and Synthetic Polymers. The addition of low concentrations of certain inexpensive, water-soluble, and high-molecular-weight organic polymers can be an effective way to significantly increase the viscosity of an aqueous solution, such as the water of a polymer flood. Biopolymers such as xanthan can be used in polymer flooding. However, synthetic hydrolyzed polyacrylamide (HPAM) has been the most widely used polymer during polymer flooding.

Biopolymers. Biopolymers can be attractive for polymer-flooding application because of their insensitivity to water salinity and hardness and their insensitivity to mechanical shear degradation. Compared to the synthetic HPAM, xanthan polymers are remarkably resistant to mechanical degradation (Seright et al. 2009). Biopolymers tend to be disfavored for their relatively high cost, sensitivity to microbial attack, and by the common occurrence of fermentation-process-induced cell debris that impairs polymer-solution injection into reservoir matrix rock. However, recent research has demonstrated that cell debris content in xanthan polymers is manufacturer dependent (a quality control issue) so that matrix-rock plugging tendency due to cell debris should not be considered an inherent limitation of xanthan polymers (Seright et al. 2009).

Synthetic Polymers. Synthetic HPAM polymer is favored due to its low cost, its commercial availability, its viscosity-enhancing power in low-salinity brines, its normally relatively good injectivity, and its resistance to microbial degradation. HPAM polymer is disfavored because of its sensitivity to water salinity and hardness and due to its sensitivity to mechanical or shear degradation. **Table 7.1** outlines the polymers commonly used in oilfield applications for conformance improvement.

7.3 Residual Oil Recovery and Permeability Reduction

7.3.1 Polymer Flooding and Residual Oil Recovery. The most important mechanism by which polymer flooding improves conformance and volumetric sweep efficiency is the viscosity-enhancement of the oil-recovery drive fluid. As such, polymer flooding would not be expected to recover any of the residual oil saturation. However, some researchers have somewhat recently suggested the possibility that the viscoelastic properties of exceptionally high-molecular-weight HPAM polymer might be able to promote, under certain situations, the recovery of some of the residual oil saturation (Demin et al. 2000; Delshad et al. 2008; Jiang et al. 2008). At the time of the writing of this book (2011), this was a controversial subject. But, at best, any viscoelastic effects occurring during polymer flooding would normally only promote the recovery of residual oil that is a second order amount as compared to the amount of oil produced by the viscosity-enhancing conformance-improvement mechanism.

TABLE 7.1—COMMON OILFIELD POLYMERS USED IN CONFORMANCE IMPROVEMENT (Sydansk 2007)	
Biopolymers	
Biopolymer Type	Characteristics
Xanthan	A widely used biopolymer for polymer waterflooding. It is derived from a microorganism fermentation process that usually leaves a substantial amount of cell debris in the final polymer solution. Flooding reservoir matrix rock with filtered xanthan polymer solutions tends to result in much less permeability reduction than the comparable flooding with appropriate synthetic acrylamide polymer solutions. Xanthan polymers and the resultant solution viscosity are relatively insensitive to the salinity and hardness of the formation brine and insensitive to mechanical shear degradation; however, they are quite susceptible to biological degradation.
Scleroglucan	This biopolymer presents a triple-stranded molecular conformation that provides the biopolymer with favorable stability and performance properties for use during high-temperature (e.g., ~195°F) polymer waterflooding.

<table>
<tr><td colspan="2" align="center">TABLE 7.1—COMMON OILFIELD POLYMERS USED IN CONFORMANCE
IMPROVEMENT (Sydansk 2007) (Cont.)</td></tr>
<tr><td colspan="2" align="center">Synthetic Polymers</td></tr>
<tr><td>Synthetic-Polymer Type</td><td>Characteristics</td></tr>
<tr><td>Polyacrylamide (PAM)</td><td>PAM tends to adsorb onto reservoir-rock surfaces, particularly onto sands and sandstone pore surfaces. PAM does not offer the viscosity-enhancing function, and it does not propagate as well through sand reservoirs as partially hydrolyzed PAMs (HPAMs). Technical-grade PAM normally contains 1 to 2 mol% carboxylate (hydrolyzed), which is enough to be an effective polymer for use in conformance-improvement polymer gels that involve chemical crosslinking reactions occurring through the polymer's carboxylate groups. PAM is normally not referred to as hydrolyzed PAM until the carboxylate content exceeds approximately 2 mol%.</td></tr>
<tr><td>HPAM</td><td>This is the most widely employed water-soluble polymer for use in both polymer waterflooding and oilfield conformance polymer-gel treatments. HPAM tends to be a better viscosity-enhancing agent in low-salinity brines and tends to adsorb less than unhydrolyzed polyacrylamides onto the rock surfaces. HPAMs are salt sensitive, so they perform better in reservoirs having low-salinity brines. For use in most polymer waterfloods, a 30% hydrolysis level seems to be optimum in terms of viscosity enhancement and in minimizing adsorption onto reservoir-rock surfaces. For use in crosslinked-polymer-gel treatments, the optimum level of hydrolysis is in the 5 to 10 mol% range because gel strength is maximized and unproductive intramolecular crosslinking is minimized. HPAMs come in two forms: the acid form or the salt form. The use of the salt form of HPAM is favored over the use of straight PAM in most polymer-flooding and polymer-gel applications. HPAMs are sensitive to high-salinity environments because the electrostatic field around the carboxylate groups induces the shrinking of the HPAM molecules, which decreases the viscosity-enhancing functionality of the HPAM solution. In high-temperature ($\geq$250°F) reservoirs, HPAMs undergo high levels of autohydrolysis and the presence of divalent cations (calcium and magnesium) induces polymer phase changes that precipitate the polymer out of the solution. Thus, the polymer solution loses most of its viscosity-enhancing function, which is the major limitation of the use of HPAMs for polymer flooding in high-temperature reservoirs.</td></tr>
</table>

7.3.2 Polymer Flooding and Permeability Reduction. There is a second mechanism of polymer flooding in reservoir matrix rock that needs to be noted. That is, as polymer-solution flows through reservoir matrix rock, it induces some permeability reduction. The degree of permeability reduction induced during polymer flooding increases with decreasing permeability of the reservoir rock and also increases with increasing molecular weight (molecule size) of the polymer.

Some professional practitioners of polymer flooding have claimed that the secondary permeability-reduction mechanism component contributes to polymer-flooding functionality in terms of helping to improve conformance. This has been a controversial contention (Sydansk 2007).

TABLE 7.1—COMMON OILFIELD POLYMERS USED IN CONFORMANCE IMPROVEMENT (Sydansk 2007) (Cont.)	
Synthetic Polymers	
Synthetic-Polymer Type	Characteristics
Copolymers containing AMPS monomers (2-acrylamido-2-methyl-propanesulfonic acid monomers) and acrylamide monomers	These types of acrylamide polymers are recommended for polymer waterflooding of high-temperature (e.g., ≥200°F) and high-salinity reservoirs, where the AMPS copolymer's performance and stability are better than the performance of comparable HPAMs.
Copolymers of vinylpyrrolidone and acrylamide; ter-polymers of vinylpyrrolidone, acrylamide, and acrylate	These polymers are candidates for polymer floods and conformance polymer-gel treatments in high-temperature reservoirs having harsh environments. The main downsides of these polymers are their high cost (as compared to conventional acrylamide polymers) and their low molecular weight.
Cationic PAMs	These acrylamide polymers have positively charged chemical groups attached to some of the polymer's pendant amide groups or acrylamide polymers that have been copolymerized with monomers containing positively charged pendant groups. These polymers have a strong tendency to adsorb onto reservoir-rock surfaces, especially sand and sandstone surfaces. Cationic acrylamide polymers are useful for specialized applications in conjunction with a variety of conformance-improvement treatments. These applications include: (1) polymer anchoring agents to help promote polymer-gel adsorption onto reservoir-rock surfaces, (2) bridging-adsorption and/or flow-induced-adsorption polymers for injection before a conformance gel treatment, and (3) polymer for use in certain polymer disproportionate-permeability-reduction conformance treatments. Water-soluble cationic acrylamide polymers are manufactured in a wide variety of forms and chemistries.
New synthetic polymeric systems: hydrophobically modified water-soluble polymers or hydrophobically associative polymers	These polymers differ from classical water-soluble polymers in that they carry low amounts of hydrophobic monomers capable of creating physical associations with each other. These polymers are characterized for exhibiting higher viscosities at low shear rates. Other properties include their unique interactions with surfaces and the ability to form physical gels at high concentrations. The steady shear rheological properties of hydrophobically modified water-soluble polymers make them particularly attractive for polymer flooding because only low concentrations of polymer are required to achieve a given mobility ratio. These polymers can be used in high-salinity reservoirs. Although untested, hydrophobically modified water-soluble polymers may be useful for water-shutoff applications, conformance control, and acid-diversion well-treatment operations. Even though the adsorption properties of these polymers are useful for well treatments, their application as mobility-control agents still needs more research because their adsorption behavior can potentially be an issue during polymer flooding (Dupuis et al. 2010).

7.4 Polymer Flooding Effects—Heterogeneous and Homogeneous Reservoirs

7.4.1 Heterogeneous Reservoirs. During polymer flooding in a heterogeneous reservoir matrix rock, the goal is to divert fluid flow from high-permeability reservoir flow paths to low-permeability reservoir flow paths in order to recover "bypassed" oil that is residing in the low-permeability reservoir matrix rock. However, it has been documented that the opposite happens (Sydansk 2007). Polymer molecules of a given molecular weight and size tend to more effectively, and to a greater degree, reduce the permeability of the smaller pore-size openings (due to adsorption and mechanical entrapment) of the lower-permeability matrix reservoir rock. Thus, the polymer-flood solution will tend to preferentially reduce the permeability of the lower permeability reservoir rock and to divert flooding fluids to flow through the higher-permeability matrix reservoir rock. So in this case, the permeability-reducing component of polymer flooding proves to be counterproductive. For this type of conformance-improvement polymer flooding, the permeability-reducing component of the polymer flood should be minimized. Otherwise, polymer flooding will reduce the flow capacity of low permeability rock by a greater factor than that of the high permeability rock, which would not aid vertical sweep efficiency, as has been quantitatively demonstrated by Seright (2010).

Still, properly managed polymer flooding is an effective technique to improve conformance and oil recovery when flooding heterogeneous matrix-rock reservoirs (permeability heterogeneity). When there is permeability variation within a matrix-rock oil reservoir, the increased viscosity of the oil-recovery drive fluid during polymer

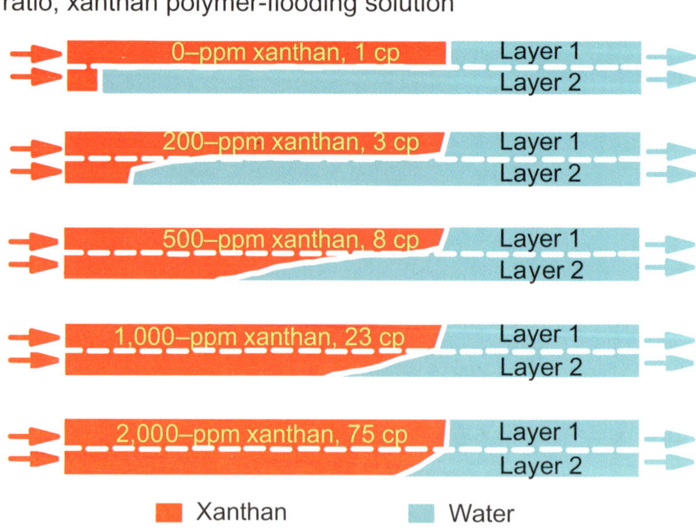

Fig. 7.1—Experimental flooding study showing how increasing the viscosity of a polymer solution improves sweep efficiency in layered porous-media strata with differing permeabilities and where crossflow is present (Sorbie and Seright 1992).

flooding improves the sweep efficiency of the flood and improves oil recovery. **Fig. 7.1** illustrates how viscosity enhancement during polymer flooding improves conformance and sweep efficiency when permeability heterogeneity is encountered in porous media.

7.4.2 Homogeneous Reservoirs. When employing polymer flooding to reduce mobility-induced viscous fingering during flooding in a "homogeneous" reservoir containing "viscous" oil, the permeability-reducing component of polymer flooding is not necessarily counterproductive. In this case, the permeability-reducing component of polymer flooding improves the mobility ratio, which in turn reduces viscous fingering and improves sweep efficiency. This is shown conceptually in **Fig. 7.2,** where the same volume of polymer solution was injected in the two cases. But, the caveat here is that most oil reservoirs are not perfectly homogeneous, and as a result, the negative impact of polymer-induced permeability reduction on polymer-flooding performance in heterogeneous reservoirs must still be taken into account during such polymer flooding.

7.4.3 Combined Mechanisms During Polymer Flooding. In view that no oil reservoir is perfectly homogeneous, then both of the above-discussed polymer-flooding mechanisms for improving conformance and sweep efficiency during a polymer flood must be simultaneously operating to help promote conformance improvement and incremental oil recovery. That is, both the viscous-fingering-mitigation mobility-ratio mechanism and the permeability-heterogeneity improved-sweep mechanism must be functioning simultaneously during any given polymer flood.

7.5 Flooding Design and Operations Considerations

7.5.1 Polymer Flooding Performance Considerations. A partial list of important considerations that can affect the performance of polymer flooding include:

- The chemical, mechanical, thermal, and biological stability of the polymer
- Polymer injectivity as it will actually occur in the wells of the field being flooded

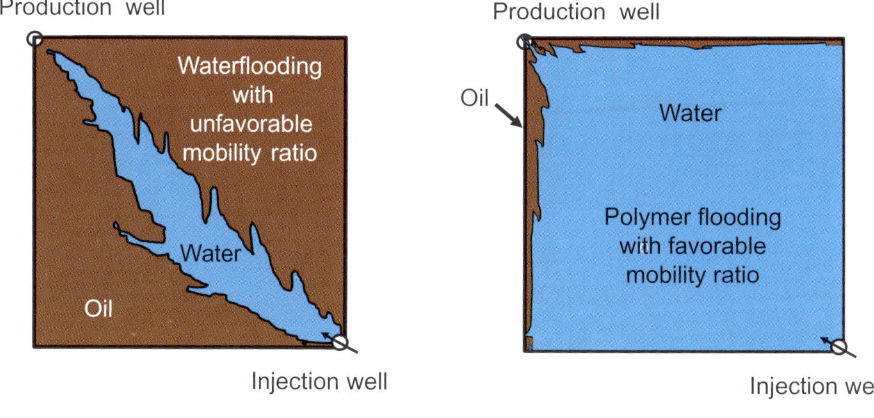

Fig. 7.2—Laboratory flooding experiment demonstrating how viscosity-enhanced polymer flooding of viscous oil can improve conformance and sweep efficiency.

- Polymer flooding volume and polymer concentration
- Polymer quality-control program during flooding
- Polymer propagation, retention, and rheological issues during flow through reservoir matrix rock
- Polymer sensitivity to salinity and hardness
- Polymer viscosity-enhancing performance as a function of rheological flow shear
- Polymer viscosity-enhancing performance as a function of polymer molecular weight and hydrodynamic size
- Possible need for polymer-solution filtration
- Favorable well completions for polymer-solution injection
- Polymer dissolution under field conditions
- Maximum injection pressures permitted during polymer-solution injection.

7.5.2 Polymer Retention—Planning Ahead. From the previous list emphasis is given to polymer retention because it constitutes an important technical and economic consideration that needs to be taken into account *prior to* initiation of a polymer-flooding project (Sydansk 2007). Retention of polymer in a reservoir can result from adsorption, entrapment, or—with improper application—physical plugging (Lyons and Plisga 2005). Retention for a given polymer during a polymer flood follows the trends listed here:

- Increases as the permeability decreases
- Increases as the polymer molecular weight increases
- Increases as the clay content in the reservoir rock increases
- Usually decreases as oil wetness increases
- Tends to increase in sands and sandstones reservoirs with decreasing anionic charge and increasing cationic charge of the polymer's pendant groups
- Has been reported to increase at times in the presence of crude oil (Sydansk 2007).

Polymer retention should be carefully determined prior to initiating a polymer waterflood. Polymer retention is best estimated by conducting flooding experiments using reservoir rock and fluids at reservoir temperature, and it is usually reported as mass of polymer adsorbed per unit volume of reservoir rock in terms of pounds of polymer adsorbed per acre-foot of reservoir, lbm/acre-ft. Polymer retention as measured during field projects has been reported to range from 20 to 400 lbm/acre-ft bulk volume, with desirable retention reported to be less than 50 lbm/acre-ft (Sydansk 2007). **Table 7.2** summarizes the main mechanisms of polymer retention in porous media.

7.5.3 Polymer Concentration and Volume. Polymer flooding with synthetic HPAM polymer has historically been favored for application in oil reservoirs possessing rather high permeabilities (>10 md, but best at >100 md), low reservoir temperatures (<210°F, but best at <170°F), and reservoirs where the polymer-solution makeup water and the reservoir water are fairly fresh (normally <9,000 ppm salinity).

Prior to approximately 1995, many conformance-improvement polymer floods did not perform as well as expected for the simple reason that not enough polymer concentration

TABLE 7.2—MAIN MECHANISMS OF POLYMER RETENTION IN POROUS MEDIA (Sydansk 2007)	
Mechanism of Polymer Retention	**Brief Description**
Polymer adsorption	Often the major cause of polymer retention during polymer floods because of physical adsorption and not from chemisorption.
Mechanical entrapment	The result of larger polymer molecules becoming lodged in narrow flow channels (e.g., pore throats). Consequences of polymer mechanical entrapment are permeability reduction, polymer retention, and the loss of the entrapped polymer's favorable viscosity-enhancing functionality throughout the remainder of the polymer flood.
Hydrodynamic retention	Results from polymer molecules becoming temporarily trapped in stagnant flow regimes by hydrodynamic drag forces. This polymer-retention mechanism is more significant in low-permeability reservoir rock, and it is considered to be a relatively small contributor to the total polymer retention during a polymer flood.
Precipitation	Polymer precipitation from solution occurs especially in the presence of high reservoir-brine salinity. Polymer precipitation is particularly problematic when conducting flooding with HPAM in high-temperature reservoirs having formation waters containing hardness-inducing divalent cations.

and volume were injected during the polymer-flooding operation (Sydansk 2007). It is now recommended that the *starting point* in the designing of a polymer flood should be to use 900 ppm polymer (or more) in the flood water and to inject at least 0.5 pore volume of polymer solution.

7.5.4 Produced-Polymer Handling for Cost Control. During any polymer flooding operation, as soon as the polymer displacing fluid breakthrough at the producing wells, the concentration of polymer in the produced fluids increases as the polymer flooding progresses. According to Demin et al. (2002a), during the polymer flooding conducted up to 2002 in the Daqing oil field (China), about 30 to 50% of the injected polymer has been produced. The maximum produced polymer concentration has been reported to be about two-thirds of the concentration of the injected polymer. The increased concentration of polymer in the produced fluids causes several operational problems. Some of these problems are listed below.

1. Corrosion at producing-well tubing and equipment. Corrosion appears to be related to polymer bonding on equipment surfaces, providing a favorable environment for corrosion pitting (Christopher et al. 1988; Demin et al. 2002a). Moffitt (1993) also reported that the production of large amounts of polymer led to increased tubing and rod corrosion rates. This problem increases the frequency of workovers and increases production costs (Yang et al. 2008). Many problems in handling polymers are caused by the negatively

charged polymer molecules that are attracted to the positively charged multivalent metal ions. To minimize corrosion, special valves, coatings, and fittings must be used (Miglicco 1986).

2. The viscoelastic nature of polymer fluids has a marked influence on injection, production, and gathering equipment and systems. The vibration in the system increases, efficiency decreases, service life is shortened, and maintenance work is greatly increased. The rheological characteristics of the polymer induce normal forces acting on the equipment—such as on beam pump sucker rods—causing pronounced eccentric mechanical wear on the side of the sucker rods that results in very short service life for beam-pump wells. These normal forces significantly lower the efficiency of centrifugal pumps, increase the vibration, and accelerates mechanical degradation in triplex pumps and in the whole system (Demin et al. 2004).

3. Decrease in production rate because of plugging of the immediate wellbore by solid fines, probably aggravated by the presence of high polymer concentrations (Christopher et al. 1988).

4. Oil/water separation becomes more complicated; it seems that at high polymer concentration, water in oil emulsions are very stable and the de-emulsification process becomes much more difficult (Feng et al. 1994).

5. Difficulties during the disposal of produced polymer solutions have been reported (Moffitt 1993). High concentration of polymer in the produced fluids causes operational problems during dehydration; for instance, electrochemical dewatering takes a longer time, increasing the electric load necessary for dewatering. Therefore, treating time and chemical dosage increase (Feng et al. 1994).

The effects of produced polymer on the total production, gathering, and fluid-treatment systems must be considered as a fundamental part of the polymer-flooding-process design and operation. The successful handling of produced polymer can substantially reduce production costs and, also, any environmental impact.

7.6 Foam Flooding

Foam flooding/injection is a special-case technology for improving conformance in oil reservoirs. Foams both increase the apparent viscosity of the foam flooding fluid and substantially reduce permeability in foam-flooding flow paths. Conformance-improvement foam flooding is discussed in Chapter 9.

Chapter 8

Improving Conformance by Reducing Permeability

8.1 Permeability-Reducing Materials for Improving Conformance

As mentioned in the previous chapter, there are two distinct conformance-improvement techniques that can be applied within the oil-reservoir rock itself in order to treat conformance problems that can only be successfully treated by material that is placed, and operates, within the reservoir rock. The first technique is to increase the viscosity of the oil-recovery drive fluid. The second technique is the placement of a permeability-reducing material in the offending reservoir high-permeability flow channels.

Conformance-improvement techniques that function by reducing the permeability of high-permeability flow paths within oil reservoirs, need to perform, at times, not only in the near- and intermediate-wellbore region of a reservoir, but also in the far-wellbore region. As such, these techniques normally involve the injection of relatively large volumes of the conformance-improvement fluids.

However, as compared to conformance-improvement techniques that involve increasing the viscosity of the oil-recovery drive fluid, conformance-improvement techniques applied to reduce the permeability of high-permeability flow paths within oil reservoirs are normally of intermediate large volumes. As a result, such conformance-improvement techniques are referred to as *treatments,* and not flooding operations. Typical volumes of a permeability-reducing conformance-improvement treatment range from 10 to 40,000 bbl. Exceptions to this treatment-volume guideline are colloidal-dispersion-gel (CDG) sweep-improvements that are discussed in Section 8.2.1 and that can range up to several 100,000 bbl in volume, and resin fluid-shutoff treatments that can be viewed as a hybrid between wellbore and reservoir near-wellbore treatments (discussed further in Section 8.2.3), in which treatment volumes can be less than 2 bbl (Sydansk 2007). The volumes of permeability-reducing conformance-improvement treatments are seldom measured in terms of reservoir or well-pattern pore volumes (as is typical for viscosity-enhancement oil-recovery flooding operations for conformance improvement). Permeability-reducing conformance-improvement treatments can be, and are, applied from both the injection-well and production-well side.

8.2 Types of Permeability-Reducing Conformance-Improvement Treatments

There are a large number and variety of different materials that can be successfully used in permeability-reducing conformance-improvement treatments that are placed in high-permeability flow channels within conventional oil reservoirs. The following subsections will briefly discuss a number of these treatment technologies.

8.2.1 Gels. Gels, in their many forms, have proven to be one of the most effective and most popular materials for use in permeability-reducing conformance-improvement treatments (Seright et al. 2003; Sydansk 2007).

Introduction to Gels. Gels are a fluid-based system to which solid-like structural properties have been imparted. In other words, classical bulk gels are fluid-based systems in which the base fluid has acquired a continuous 3D solid-like structure throughout the gel mass. These structural properties are often elastic in nature. All oilfield conformance-improvement gels to be discussed in this chapter are *aqueous-based gels.* In the modern oilfield and technical literature, the term "polymer gel," as is used in this chapter, includes the elastic and semi-solid materials that result from chemically crosslinking together water-soluble polymers in an aqueous solution.

For the most part, conformance-improvement classical bulk gels are injected into the reservoir as a "watery" gelant (pre-gel) solution, or as a partially formed gel; after the gelation- onset time for the particular gel at reservoir conditions, the gelant solution (or partially formed gel) matures and sets up in the reservoir. That is, the "watery" (or near watery) solution matures into a gel that has solid-like properties.

Aqueous *polymer gels* are formed when relatively low concentrations of a water-soluble high-molecular-weight polymer is dissolved in water (e.g., produced water) and a chemical crosslinking agent, which is incorporated into the gel formulation, will chemically crosslink the polymer molecules together while the polymer molecules remain dissolved in the aqueous solution. For classical bulk polymer gels, the continuous chemically crosslinked polymer network adds elastic solid-like properties to the resultant gel. The water (the majority of the gel's makeup) is encapsulated within the interstices of the 3D crosslinked-polymer network. The resultant aqueous polymer gel is a functionally-effective and cost-effective agent for use as a permeability-reducing and/or fluid-flow-blocking agent in conformance-improvement treatments.

Gel Treatment Concentrations. Oilfield crosslinked-polymer gels can possess rigidity up to and exceeding that of Buna rubber. Conformance-improvement polymer gels contain polymer concentrations in the range of 1,500 to 100,000 ppm, but more commonly 2,000 to 50,000 ppm, and most frequently in the range of 3,000 to 12,000 ppm. Conformance-improvement polymer gels are often formulated with relatively inexpensive commodity polymers.

Broad Application of Gels in the Oil Industry. Polymer gels have found broad application in the oil industry as highly effective permeability-reducing and/or blocking agents for use in conformance-improvement treatments. This has resulted in part because gels are often an exceptionally cost-effective blocking and permeability-reducing agent for use in a number of oilfield conformance-improvement treatments, and such gels are, for the most part, nothing but water (often produced brine in the case of oilfield gels), with the remainder of the gel constituents being "low" concentrations of relatively inexpensive polymers and chemical crosslinking agents.

Bulk Gels and Microgels. Conformance-improvement polymer gels can occur in two distinctive forms: *bulk gels* and *microgels*. Bulk polymer gels are *classical gels* where the crosslinked polymer network is essentially infinite in nature. A bulk polymer gel, when formed in a beaker, is characterized as a gel that has a crosslinked polymer network extending throughout the entire gel volume in the beaker. Microgels are gels that are formed at low polymer concentrations and at polymer concentrations below the polymer's critical overlap concentration in the gel's makeup brine. Microgel particles are tiny particles and are not normally discernable to the naked eye. Polymer microgels are dominated by intramolecular crosslinking (within the same polymer molecule), while polymer bulk gels are dominated by intermolecular crosslinking involving different polymer molecules.

Oilfield conformance-improvement gels include not only polymer gels, but also inorganic gels, preformed particle gels, and monomer gels—all of which are briefly reviewed in this chapter.

Gel Treatment Applications. Oilfield gel conformance-improvement treatments can be applied in a number of different treatment forms, with the following being a partial list:

- Sweep-improvement treatments
- Water-shutoff treatments
- Gas-shutoff treatments
- Zone-abandonment treatments
- Squeeze and recompletion treatments
- Water- and gas-coning treatments involving fracture and other high-permeability-anomaly coning problems

Benefits of Gel Treatment Applications. A partial list of conformance-improvement gel treatment benefits is as follows.

- Generate incremental oil production through conformance improvement
  - Possibly generate large volumes of incremental oil production per unit cost or weight of treatment chemicals used
- Substantially reduce oil-production operating expenses per unit cost or weight of treatment chemicals used
- Reduce competing water production that can be unproductive, costly, and, at times, environmentally unfriendly
- Reduce competing gas production that can be unproductive, excessive, and economically detrimental
- Improve the performance of an ongoing oil-recovery operation
- Reduce certain environmental liabilities by reducing the amount of excessive and unnecessary coproduction of environmentally unfriendly fluids, such as high-saline and high-hardness reservoir brines
- Extend the economic lives of marginal wells, well patterns, and oil fields

Applicability and Characteristics of an Ideal Gel Technology. The applicability and characteristics of an "ideal" oilfield conformance-improvement gel technology are as follows.

- Applicable as:
  - Injection and production well treatments
  - Sweep improvement treatments
  - Water and gas shutoff treatments
- Applicable to:
  - All reservoir mineralogies and lithologies
  - A wide variety of reservoir and flooding-operation conformance problems
  - A broad range of reservoir temperatures
  - Over a broad injection- and formation-water pH range
- Characteristics:
  - Be a single-fluid system
  - Possess a robust gel chemistry
    - Be insensitive to oilfield and reservoir environments and chemical inter-ferences (especially H_2S and CO_2)
    - Be insensitive to all reservoir minerals and fluids
  - Involve a simple and straightforward gel-forming chemical system
  - Be stable and functional in the reservoir over the long term (i.e., years)
  - Provide for a broad range of
    - Gel strengths, including rigid gels
    - Highly controllable and predictable gelation-onset delay times
  - For matrix-rock reservoir treatments, have a gel version that is readily inject-able into reservoir matrix rock
  - For fracture-problem treatments, have a gel version that is only injectable and placed within the fractures
  - Be environmentally acceptable and friendly
  - Be formulated with low concentrations of relatively inexpensive chemicals
  - Be formulated with readily available (preferably commodity) chemicals
  - Reduce, in porous matrix rock, the permeability to water flow more than the permeability to oil and gas flow

Types of Oilfield Conformance-Improvement Gel Technologies. **Table 8.1** con-tains a list of various types of oilfield conformance-improvement gel technologies. This list is not intended to be an exhaustive list of such gel technologies. A selected number of the conformance-improvement gel technologies presented in Table 8.1 are briefly discussed as follows.

Silicate Gels. Silicate gels have been the most widely applied *inorganic* confor-mance-improvement gel technology in past years. Silicate gels are likely the oldest conformance-improvement gel technology. The silicate portion of the gel technology itself is environmentally friendly. Whether the acid material required to gel the silicate solution is environmentally friendly needs to be evaluated for each acid material employed. At the present time (2011), silicate-gel conformance-improvement treat-ments are not being widely applied. Some silicate gelant solutions have experienced injectivity problems during injection into reservoir matrix rock. Silicate gels are some-what brittle gels and are not particularly strong gels (as compared to polymer gels). A challenge for silicate gels is that, for any silicate-gel formulation, as the gelation-onset time is increased, there is a tendency for the gel strength to be concurrently reduced. Additional-ly, more studies should be conducted on the long-term stability and performance of

TABLE 8.1—GELS FOR USE IN CONFORMANCE-IMPROVEMENT TREATMENTS

- ➢ Inorganic based (bulk gels)
  - ▪ Silicate gels
  - ▪ Aluminum-based gels
- ➢ Organic-based polymers
  - ▪ Bulk gels
    - o Synthetic or biopolymers
      - ♦ Acrylamide polymers (most widely used polymer)
      - ♦ Xanthan biopolymer
    - o Organic crosslinkers
      - ♦ Aldehydes
        - ▫ Phenol-formaldehyde and derivatives
      - ♦ Polyethyleneimine
    - o Inorganic crosslinkers
      - ♦ Al(III) based
      - ♦ Zr(IV) based
      - ♦ Cr based
        - ▫ Cr(VI) redox
        - ▫ Cr(III) with inorganic anions
        - ▫ Cr(III) with organic carboxylate complex ions
  - ▪ Monomer gels (organic-monomer-based in-situ polymerization)
    - o Acrylamide monomer
    - o Acrylate monomer
    - o Phenolics
  - ▪ Lignosulfonate gels
  - ▪ Preformed particle gels
    - o Swelling organic-polymer "macroparticle" gels
- ➢ Mixed silicate and acrylamide-polymer gels
- ➢ Microgels
  - ▪ Microgels with narrow particle-size distribution
  - ▪ CDGs
    - o Aluminum-citrate crosslinked
    - o Chromic-triacetate crosslinked
  - ▪ Delayed "popping"/swelling microgels (Bright Water^{TM})

conformance-improvement silicate gels and should be conducted over a broad range of reservoir conditions (Sydansk 2007).

Polymer Gels. Polymer gels, involving the use of acrylamide polymers, have evolved to become the most widely applied conformance-improvement gel technology. For application to oil reservoirs with high temperatures ($\geq$250°F), the use of organic crosslinking agents may be favored.

Chromium(III)-Carboxylate/Acrylamide-Polymer (CC/AP) Gels. CC/AP gels have emerged to become a popular and widely applied polymer-gel technology for application as a conformance-improvement treatment—including sweep-improvement treatments and water- and gas-shutoff treatments. These are aqueous acrylamide-polymer gels in which the chemical crosslinking agent is a chromium(III) carboxylate complex. CC/AP gels are characterized as having exceptionally robust gel chemistry, being highly insensitive to oilfield and reservoir interferences and environments, and being applicable over an exceptionally broad pH range. As a result, these gels, when properly formulated, are applicable to the acidic conditions associated with CO_2 flooding—where most of the earlier oilfield polymer gels did not function well due to the acidic conditions of CO_2 flooding. Chromium(III), as found in the crosslinking agent of this gel technology, is relatively non-toxic, but it is highly regulated. This single-fluid gel technology provides for a wide range of gel strengths and a wide range of controllable gelation-onset delay times. CC/AP is applicable over a broad range of reservoir temperatures. However, the upper reservoir temperature limit for applying the total-shutoff version of CC/AP gels in matrix rock near-wellbore is reported to be 300°F (Sydansk and Southwell 2000).

CC/AP gels are applicable to an extensive range of conformance problems and reservoir mineralogies and lithologies. Chromic triacetate ($CrAc_3$) is the often-preferred crosslinking agent used in conjunction with the CC/AP gel technology. A chemical gelation-rate-acceleration additive package involving chromic trichloride has been developed for use with the CC/AP gel technology. Three chemical methods are available to retard the rate of gelation of CC/AP gels that are applied to high-temperature reservoirs. These methods include the use of gelation-rate retardation agents such as strong carboxylate ligands (e.g., lactate), the use of ultra-low-hydrolysis polyacrylamides within the gel formulation, and the use of low molecular weight (MW) acrylamide polymers (Sydansk 2007; Sydansk and Southwell 2000; Sydansk 1993; Sydansk 1990).

Monomer Gels. Monomer gels are a form of conformance-improvement gels. These gels are based on the in-situ polymerization of organic monomers to form polymer gels in situ. These gels, which may be formulated with and without the inclusion of crosslinking monomers, have been developed and successfully applied during oilfield conformance-improvement operations (Sydansk 2007).

Early monomer-gel treatments were often based on the in-situ polymerization of acrylamide monomers, but this is seldom practiced presently (2011) due to toxicity and environmental concerns. Most recent conformance-improvement monomer-gel technologies for oilfield application are based on the in-situ polymerization of relatively less toxic acrylate monomers. An advantage of monomer gel technologies is the low water-like viscosity of the gelant solution. Disadvantages of these monomer gels include cost issues and the sensitivity of the polymerization reaction to oilfield interference and environments. Care is also required to distinguish between "linear" (uncrosslinked) and crosslinked oilfield monomer gels.

An older monomer gel technology involved gels formed from the reaction of formaldehyde with phenol. A current concern with this gel technology is the toxicity and the environmental issues associated with the use and handling in the field of formaldehyde and phenol and/or their chemical derivatives.

Microgels. Microgels for conformance-improvement-treatment applications are used in a variety of forms. One form of *discontinuous* microgels has been studied somewhat extensively by European researchers (Chauveteau et al. 2003; Chauveteau et al. 2001). These low-polymer-concentration acrylamide-polymer gels, which are formed above ground prior to injection, reportedly have an unusually narrow particle-size distribution.

Another form of conformance-improvement microgel technology that is emerging involves the injection of nano-sized crosslinked-polymer particles. These particles are designed to pop/swell-in-size by a factor of roughly 10 after being placed "deep" in the reservoir and after experiencing some gel-particle-popping trigger, usually increased reservoir temperature (Frampton et al. 2004; Pritchett et al. 2003). These *delayed "popping"/swelling microgels* (Bright Water™) are systems of time-delayed, highly expandable thermal sensitive particles that can be used to cause blocking effect at the temperature transition in an oil reservoir, improving the sweep efficiency of a water flood. The material is a highly crosslinked, sulfonate-containing polyacrylamide microparticle in which the conformation is constrained by both labile and stable internal crosslinks. When subject to elevated temperatures, the rate of decrosslinking of the labile crosslinker accelerates. This reduces the crosslink density of the particle and allows the particle to expand by absorbing the surrounding water. The particles are applied in the constrained state—called *kernel particles.* After heating, these are able to swell, adopting a much expanded configuration called *popcorn* (Frampton et al. 2004).

The reduction in reversible crosslink density is also time dependent and can be affected by the pH of the fluid. Different labile crosslinkers have different rates of bond cleavage at different temperatures. The temperature required and the mechanism of bond dissociation depends on the chemical structure of the crosslinker. For example, when the labile crosslinker is a diacrylate ester, hydrolysis of the ester linkage becomes the most likely mechanism of decrosslinking. Particles with different activation temperature can be produced if the appropriate crosslinker is used. The presence of a stable crosslinker gives conformational integrity to particles, especially after popping. The kernel microparticles are prepared using an inverse-emulsion-polymerization process to assure a preselected particle size range. Depending on the synthetic method, the original particle diameter of the polymeric microparticle can be made ranging from approximately 0.1 to approximately 3 µm. The microparticles prepared in inverse emulsion form can be dispersed into aqueous media with agitation and the aid of inverting surfactants. Due to their highly crosslinked nature, the size of the kernel microparticles changes very little in solutions of different salinity. Consequently, the rheological properties of the carrier fluid are not affected by the salinity change encountered in a subterranean formation. Accordingly, no special carrier fluid is needed for treatment. Only after the particles encounter conditions sufficient to reduce the crosslink density is the fluid rheology changed to achieve the desired effect, which is a viscosity and/or a residual resistance factor capable of diverting water out of a thief zone (Frampton et al. 2004).

A commercial trial of the delayed popping/swelling microgels (Bright Water™) was conducted in 2004 at the BP Milne Point field, in the North Slope of Alaska. The well pattern (one injection well and two producer wells) chosen for this trial had a significant estimated quantity of bypassed oil resulting in a relatively low recovery factor of 20% and a water cut of 90%. An injection survey, a chemical tracers application (Fluorescein and Rhodamine fluorescent dyes), and an interference test conducted previously in the pattern diagnosed the existence of a high-permeability connection between the injection well and one of the producer wells. The objective of the treatment was to reduce water cycling through the thief zone between the injector and producer well to achieve an increase in pattern reserves by improving sweep efficiency. The field treatment was applied by pumping into the injection well 15,587 gal (60.8 tonnes) of the particulate product at a concentration of 33,000 ppm active particles with 8,060 gal (30.4 tonnes) of dispersing surfactant over a period of 21 days in July 2004. The particulate product was mixed at 1% by volume as supplied and the surfactant at 0.5% by volume as supplied. No change in well injectivity was noted while pumping; changes in well injectivity were observed after 9 months post injection. After 11 months post injection the water cut started to decrease and oil rate to increase, totaling over 60,000 bbl of incremental oil production to the end of 2007 (Ohms et al. 2009).

Colloidal Dispersion Gels (CDGs). CDGs are a conformance-improvement technology that has been somewhat widely applied and has been heavily promoted (Wang et al. 2008a; Sydansk 2007; Smith 1995; Mack and Smith 1994). The oilfield CDG technology involves the injection of large volumes (up to 100,000s of bbl) of microgel treating fluid. CDG microgels are injected from the injection-well side for the purpose of improving conformance and waterflood sweep efficiency deep (far wellbore) in heterogeneous "matrix-rock" of sandstone reservoirs. CDG aqueous microgels are formed by crosslinking 150 to 1,200 ppm high-molecular-weight hydrolyzed-poly-acrylamide polymer with aluminum citrate or chromic triacetate. Aluminum-citrate CDG microgels are generally considered to be environmentally friendly. The application of the more classical CDG gels, in which aluminum citrate is used as the cross-linking agent, appears to be limited to where the gels can be formulated with relatively fresh and non-hardness-containing water.

For a number of reasons, the CDG conformance-improvement gel technology has proved to be controversial (Sydansk 2007; Wang et al. 2008a; Chang et al. 2006; Seright 2006). One of the controversial issues is the actual conformance-improvement mechanism by which CDG microgels function. The proponents of the CDG microgel technology claim that the discontinuous CDG microgel particles are placed in and function as permeability-reducing agents deeply (far wellbore) in the matrix-rock sandstone reservoirs of normal permeabilities (<1,000 md). However, this functionality has been argued because of the size of the relatively rapidly gelling rigid CDG microgel particles (possibly on the order of 25 μm in diameter or larger) is too large to be propagated through, and placed deeply into, matrix-rock reservoirs of normal permeabilities.

It has also been questioned whether or not the low-aluminum-crosslinker concentration is not removed by retention before crosslinking. In particular, after looking at the field, laboratory, and theoretical basis associated with CDGs, some researchers and practitioners have concluded that historically the CDG treatments have often provided little to no benefit beyond that given by a polymer flood that did not include the aluminum-citrate crosslinker—that is, the crosslinker added no value for most of these field projects (Wang et al. 2008a).

An alternative mechanism that could explain how CDG conformance-improvement microgel particles (25 μm in diameter or larger) might be functioning, is that these gel particles are being selectively placed and functioning in intermediate-high-permeability flow channels (e.g., micro fractures) that have flow apertures on the order of 100 μm.

Nanotechnology Enhanced-Oil-Recovery Opportunities. In the field of conformance-improvement technologies, Fletcher et al. (2010) recently highlighted the potential that nanotechnology offers to transform EOR mechanisms and processes. The physics and chemistry principles supporting the mobilization of oil droplets to form an oil bank, together with the engineering that ensures that the oil bank remains mobile, are underpinned by nanotechnology together with transient phenomena. Common processes that take place in EOR such as adsorption, interfacial tensions, phase behavior, contact angles, thin films, and wettability alteration occur at the nanopore scale. Thus, nanotechnology has a unique opportunity for transforming the design and execution of chemical EOR. For instance, Fletcher et al. (2010) illustrated the concept of polymer flooding for in-depth profile modification through the application of a new nanotechnology-EOR approach.

Linked Polymer System. In the field of nanotechnology with applications to reservoir conformance-improvement, another example is the *linked polymer system* (LPS), which has recently been under development. LPS is a form of acrylamide-polymer/aluminum-citrate microgels having particle diameters in the nanometer range (Spildo et al. 2008).

Profitability, Trends, and Limitations of Gel Technology. As previously indicated, if there is the appropriate match between a given conformance problem and a particular gel technology, a noteworthy feature and advantage of oilfield gel conformance-improvement treatments is that these gel treatments can generate, in a profitable manner, relatively large volumes of incremental oil production, and/or substantial reductions in oil-production operating costs via the shutting off of the production of excessive, deleterious, and competing coproduction of nonoil fluids, such as water and gas. Gel conformance treatments are usually applied in the course of normal and ongoing primary, secondary, or tertiary oil-recovery operations. This is done in a conceptually similar manner as in the application of an acid or scale-inhibition treatment. Gel conformance treatments are an emerging oilfield technology that can help extend the life of a number of maturing oil reservoirs that are approaching their economic limit and abandonment.

An interesting trend has emerged as it relates to the application of fracture-problem conformance-improvement polymer-gel treatments. For these gel treatments that have been applied to date (2011) if there is a good match between the gel technology being employed and the conformance problem being treated, the amount of incremental oil and the profitability generated by the gel treatment generally tends to increase with increasing the volume of the gel placed into the offending zone (Sydansk 2007; Willhite and Pancake 2008).

A current limitation (2011) for available commercial conformance-improvement gel technologies is their lack of appropriate thermal stability and long-term performance when placed in high-temperature oil reservoirs (> 270°F).

Gels for High-Permeability Anomalies and High-Permeability Matrix-Rock Flow Paths. A critical aspect to the successful application of conformance-improvement bulk polymer-gel treatments is to know, determine, or deduce *before applying the gel treatment* if the conformance problem to be treated involves a high-permeability irregularity (e.g., fractures) or a high-permeability flow path within the matrix reservoir rock. This is because there are often separate versions of specific gel technologies for treating a high-permeability-anomaly vs. a high-permeability-matrix-rock flow-path conformance problem. In this respect, if an operator picks a high-permeability-anomaly gel treatment for application to a high-permeability-matrix-rock flow-path conformance problem (or vice versa) the treatment is very likely doomed to failure.

Polymer gels for treating high-permeability-matrix-rock flow-path conformance problems are designed such that upon injection and placement, the pregel solution can readily be injected into, and flow through, reservoir matrix rock of normal permeabilities (<1,000 md) in order to be placed in the desired matrix-rock volume. In contrast, polymer gels for treating high-permeability-anomaly conformance problems are designed so that the partially formed (or conceivably fully formed) gel solution upon selective injection into, and placement in the offending high-permeability anomalies (e.g., fractures) does not, and cannot, invade and damage reservoir matrix rock (to any meaningful counterproductive degree).

Gel Treatments and EOR CO_2 Flooding—An Opportunity. Currently (2011), there is a very substantial opportunity for the successful and profitable application of conformance-improvement high-permeability-anomaly-blocking polymer gels in conjunction with CO_2 IOR/EOR flooding operations that are being conducted in naturally fractured reservoirs. In the United States, CO_2 flooding has historically been applied most often to naturally fractured reservoirs. Mobility-control conformance problems are extremely acute when high-mobility CO_2 gas is flooded in naturally fractured reservoirs. At present, there is a great potential that CO_2 flooding in the United States and elsewhere could dramatically increase as the world will likely turn to sequestration of CO_2 greenhouse gas. If this comes to fruition, then subsidized CO_2 delivery to oilfield wellheads may become a reality. It would make sense to reap the synergies of CO_2 sequestration and oilfield CO_2 IOR/EOR flooding.

8.2.2 Relative-Permeability-Modification (RPM) Water-Shutoff Treatments. Certain water-soluble polymers alone and certain relatively weak polymer gels have been, and can be, used in relative-permeability-modification (RPM) conformance-improvement water-shutoff treatments that are applied to production wells of matrix-rock oil reservoirs. RPM, also termed disproportionate permeability reduction (DPR), refers to the property of these polymers and weak gels to reducing the permeability to water flow to a greater degree than to oil and gas flow in the reservoir volume in which the RPM treatment material is placed (Sydansk and Seright 2007). Several mechanisms for RPM have been proposed. These mechanisms include changes in porous medium wettability, lubrication effects, segregation of flow pathways, gel dehydration, and gel displacement, among others. Although there is no agreement upon a definite mechanism or mechanisms responsible for RPM, presently the most accepted mechanism seems to be *gel dehydration.* Therefore, this topic remains a subject under active investigation.

Seright et al. (2006) describe possible mechanisms by which pore-filling chromium(III)-carboxylate/acrylamide-polymer gels impart RPM in porous media. In this study, x-ray computed microtomography was used to indirectly visualize the RPM effect by Cr(III)-acetate-HPAM on Berea sandstone having a permeability of 0.47 darcies. A review of these mechanisms considers two conditions. First, assume a situation immediately after gel placement and gelation in a porous media, in which the water-base gel occupies basically all of the aqueous pore space and residual oil may be trapped in pore centers in a water-wet rock, as shown in **Fig. 8.1.** If water or brine is injected, an extremely high resistance to water flow is observed. This significant reduction of permeability to water is explained by the fact that the water must flow through the gel itself, which has an inherent permeability to water in the microdarcy range.

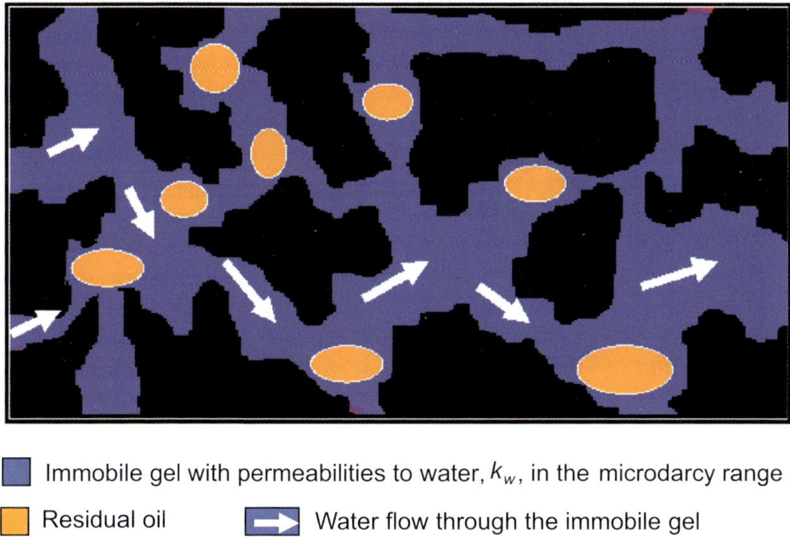

Fig. 8.1—Gel RPM effect on water flow [adapted from Seright et al. (2006)].

Therefore, the flow of water experiences a significant flow reduction through the gel-invaded water zones.

Now, consider the second condition, in which the well is returned to production after gel placement and gelation, and in which oil is typically the first fluid to contact the gel-treated region, as illustrated in **Fig. 8.2.** During oil flow, at a fixed pressure gradient, experimental observations indicate that oil flow opens up pathways along the semi-solid gel phase. Hence, the permeability to oil is restored, followed by the re-establishment of oil flow across the treated region. These pathways in the gel phase are created by the *squeezing* or *dehydration* of the gel due to the pressure gradient imposed during oil flow. *Gel dehydration* is defined as the process of removing water from the gel by imposing a pressure gradient on the gel (Seright et al. 2006). As a result, the gel must become more concentrated and its volume reduced. These suggested mechanisms of Cr(III)-acetate-HPAM gels' RPM behavior have also been observed in oil-wet porous media (Seright et al. 2006).

Willhite et al. (2002) observed that in brine displacement experiments conducted in Berea sandstone cores containing initial oil saturation after the placement of chromium acetate-polyacrylamide gels, the initial brine permeability was reduced by factors of 100 to 1,000 more than the oil permeability. When residual oil saturation is initially present as the gel is placed, the residual oil is encapsulated by the gel as shown in **Fig. 8.3a.** If the treated interval is subjected to a pressure gradient by injecting oil or by producing oil from that interval under a controlled drawdown, some oil permeability is restored due to the creation of flow paths (Fig. 8.3b). Some residual oil ganglia are reconnected as the injected oil generates the new pore space. If water is injected into the porous medium after the creation of the new pore space, the residual oil saturation in the new pore space creates substantial flow resistance to water in the new flow paths and is a major contributor to RPM, as illustrated in Fig. 8.3c. Furthermore, if oil is injected after the first waterflooding, as shown in Fig. 8.3d, the residual water exists

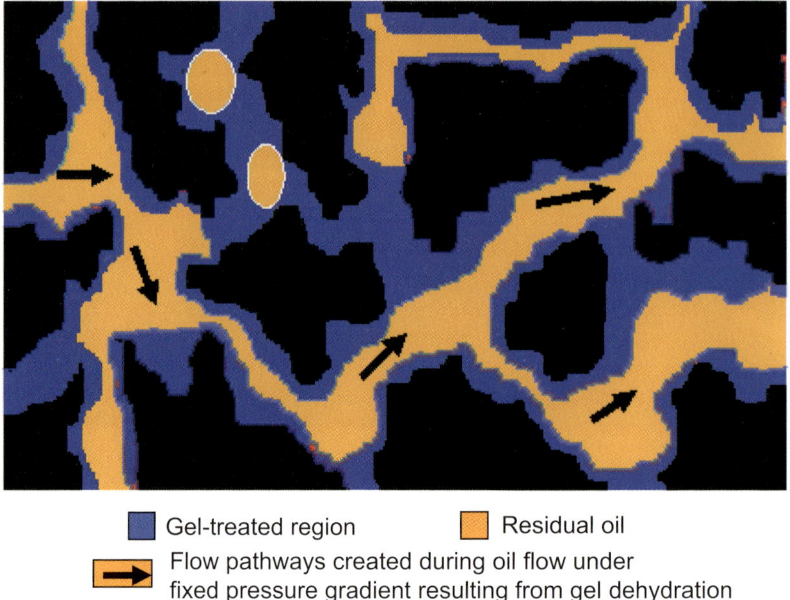

Fig. 8.2—Creation of pathways through the gel-treated region due to gel dehydration during oil flow under pressure gradient [adapted from Seright et al. (2006)].

in isolated pockets and in thin films, causing slight flow resistance to oil. Thus, oil permeability is significantly higher than water permeability because oil flows through most of the new pore space.

RPM water-shutoff treatments, especially for offshore application, are quite attractive because they can be injected and placed in a matrix-rock reservoir without having to use expensive and operationally demanding mechanical zone isolation and the associated well-workover operations.

At first glance, RPM water-shutoff treatments sound like a seductively attractive water-shutoff proposition. However, the field track record for such water-shutoff treatment has been mixed, at best, to poor. It seems that the successful application of RPM water-shutoff treatments for the purpose of rendering *long-term* water-shutoff in matrix-rock reservoirs is limited to only the excessive water-production problem where the reservoir oil-producing and water-producing strata are not in vertical pressure and fluid-flow communication, and the oil-producing strata must be producing at 100% oil cut. More information on the application of RPM water-shutoff treatments can be found in Sydansk and Seright (2007).

8.2.3 Resins. Resins are permeability-reducing conformance-improvement materials that are placed very near wellbore or in the outer portion of a wellbore (e.g., perforation tunnels or a gravel pack) for the purpose of imparting total fluid shutoff and total reduction in permeability and fluid-flow capacity (Sydansk 2007). Resins are organic polymer-based *plastic* solid materials that do not contain a significant amount of a solvent phase (e.g., as is found in a gel) and that are placed downhole in a liquid monomeric (or oligomeric) state and are polymerized in situ to the mature

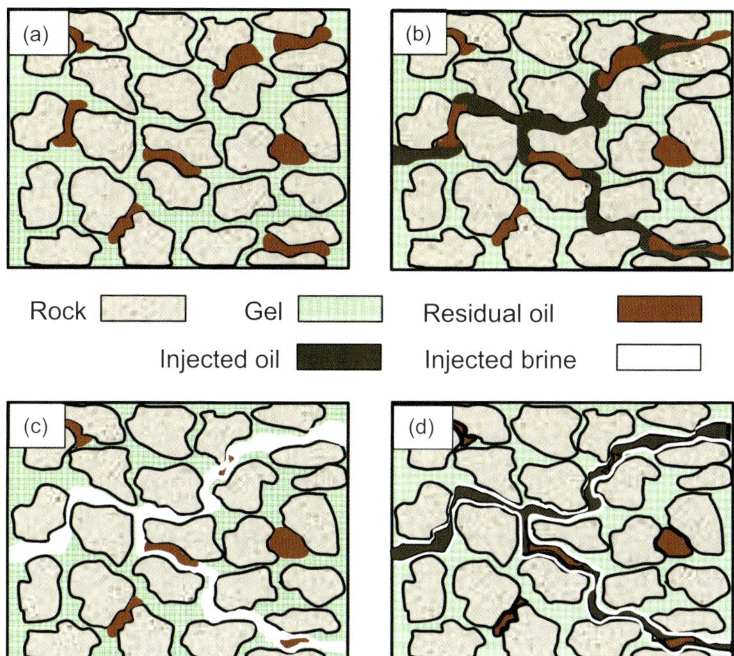

Fig. 8.3—(a) Encapsulation of waterflood residual oil flowing in-situ of chrome-acetate-polyacrylamide gelant; (b) generation of new pore space when the gel dehydrates through injection of oil; (c) trapping of residual oil in new pore space during brine flood, leading to DPR of brine; (d) flow paths of oil through new pore space, trapping low saturations of brine [adapted from Willhite et al. (2002)].

solid resin state. The use of resins for conformance improvement dates back at least to 1922.

Resins are exceptionally strong materials for use in blocking and plugging fluid flow in the wellbore and/or the very near-wellbore region. Resins for conformance-improvement fluid-shutoff purposes during squeeze treatments can normally only be placed in the wellbore, perforations, gravel packs, and other near-wellbore multidarcy flow channels. Resins have exceptionally good compressive strengths. When resins contact oil-free surfaces of reservoir rock, they usually have superior bonding strength to the rock surfaces. The vast majority of the resin field applications involving the use of the conformance-improvement resins technologies have utilized treatment volumes on the order of only 1 to 5 bbl. Resin conformance-improvement treatments are expensive on a unit volume basis. A barrel of fluid-shutoff resin can cost thousands of dollars.

Three types of resin chemistries have been somewhat widely applied for use as conformance-improvement treatments. These three types are based on, respectively, epoxy, phenolic, and furan chemistries. All three of these resins are thermosetting plastics. All three resins have good mechanical, bonding, and compressive strengths. None of these resins are highly sensitive to the pH in the wellbore during resin placement or sensitive to H_2S. These conformance-improvement resins are not substantially degraded by conventional acid treatments that are performed subsequent to the resin-treatment placement downhole. These three resins are more stable to acid than conventional Portland cement. Immature resin solutions, as they are being placed downhole, are quite viscous (often qualitatively described as having roughly the consistency of molasses).

Conformance-improvement resin treatments have been used in conjunction with waterflooding, CO_2 flooding, and steamflooding to shut off unwanted water, steam, or hydrocarbon-gas production. Because of the excellent thermal stability of phenolic and furan resins, these two resins have been successfully used during steamflooding of oil reservoirs.

Due to the small-volume nature of resin treatments, such treatments were historically often placed downhole using wireline dump bailers. More recently, the majority of fluid-shutoff resin treatments has been placed through clean and uncorroded injection or coiled tubing. The use of injection or coiled tubing for placement of resins favors the advantageous faster placement of the resin and permits the resin to be better squeezed away outside of the inner wellbore using high-differential-pressure placement.

Advantages and Disadvantages of Resins. Advantages and disadvantages of the use of resins as small-volume fluid-shutoff treatments for treating oilfield conformance problems are as follows.

Advantages. Resins possess superior:

- Mechanical strength
- Bonding strength
- Thermal stability
- Chemical inertness (e.g., the rock formation can be acidized over resins)

Disadvantages. Resin conformance-improvement treatments are:

- Constrained by limited resin penetration distances into the formation (due to a combination of high-viscosity and cost issues)
- Constrained by placement techniques and issues
- Costly on a unit volume basis
- In practice, usually limited to shallow reservoir applications
- Considered to be a niche conformance-improvement-treatment technology
- Resin application is operationally and chemically complicated

Resin fluid-shutoff treatments were somewhat widely applied a number of years ago and were relatively popular then. However, presently (2011), relatively few resin conformance-improvement treatments are being applied. Resin conformance-improvement treatments are presently only available from a few service companies.

8.2.4 Solid Particulates. Another class of materials, which are applied within the oil reservoir itself during conformance-improvement treatments and applied as permeability-reducing agents, are rigid or semi-rigid solid particles.

Classical Rigid Solids. Classical rigid solids, such as silica flour, fine sand particles, and clays, have been used (but in general not too successfully) as permeability-reducing agents in conformance-improvement treatments.

A substantial challenge to the successful use of natural solid particles as permeability-reducing agents in matrix reservoir rock is that the particle size distribution of natural solids particles (e.g., sand particles) is normally relatively large (Seright and Liang 1995). This renders such solid particles to be very difficult to place in reservoir matrix rock in a controlled, predictable, and successful manner. The pore-throat size distribution in reservoir rock also is normally relatively large. Thus, to find a solid particle material that can, and will, only plug reservoir flow channels having a certain permeability (or permeability range) and a specific average pore-throat size is a real technical challenge—and the field track record in this regard is not impressive. Furthermore, it is important to keep in mind that permeability of reservoir matrix rock is not only a function of the size of the pore throats, but also a function of the number of pore flow channels per unit area. Thus, for porous and permeable rock having a fixed pore-throat size (or pore size distribution), as the number of pore flow channels per unit area increases, the permeability of the rock also increases.

Clay-Based Slurries. In the past, the use of *clay-based* conformance-improvement treatments had received quite a bit of attention and use in Russia and China (Liu et al. 2006). Also, "gunk" clay/oil slurries have been used at times in treating high-permeability-anomaly conformance problems.

Preformed Particle Gel Technologies. Preformed particle gels are permeability-reducing agents that presently (2011) are gaining attention and popularity for use in conformance-improvement treatments. Preformed particle gels are gel particles formed above ground prior to injection. After injection and placement in the reservoir, preformed particle gels will swell in size reducing the permeability of the reservoir flow channels where they are placed. Conformance-improvement techniques based on the injection of microgels of hydrolyzed polyacrylamide crosslinked with aluminum citrate are an exception in that these microgels do not swell noticeably.

Two largely separate forms of preformed-particle-gel conformance-improvement technologies for treating high-permeability-anomaly (e.g., fracture) conformance problems have been developed and have been somewhat widely applied in the field setting. In both of these cases, the injected gel particles, as originally developed, had a diameter on the order of 1 mm or larger. For both of these preformed-particle-gel technologies, it is reported that smaller size gel particles are being developed and applied in certain instances. Preformed-particle-gel conformance-improvement treatments have been applied from both the injection-well and production-well sides.

The first of these preformed-particle-gel conformance-technologies (Bai et al. 2007; Bai et al. 2008) appears to have been developed and initially applied in China. This preformed-particle-gel technology involves aqueous gel particles that are formed by polymerizing the acrylamide monomer in the presence of a specialty crosslinking acrylamide monomer (N,N′-methylenebisacrylamide). The resultant bulk gel is then disaggregated into the preformed gel particles.

The second of the preformed-particle-gel conformance-improvement technologies was primarily developed in the United States (Abbasy et al. 2008). This preformed-particle-gel technology involves swellable aqueous-gel particles that are manufactured gel particles and gel particles involving crosslinked specialty sodium-acrylate-based polymers.

Microgels. Microgels are another important class of solid-particle-like conformance-improvement technologies that perform by the permeability-reduction mechanism. Microgel conformance-improvement technologies are often intended to be placed in, and function, deeply (far wellbore) in matrix rock reservoirs. Such microgel technologies include narrow-particle- width-size microgels, CDGs, and preformed nanosized crosslinked-polymer gel particles that pop/swell-in-size after being placed in an oil reservoir. All three of these conformance-improvement microgel technologies are discussed in more detail in Section 8.2.1.

Absolute Particle-Size Issues. Absolute particle-size issues are an important consideration when applying particulate material in conformance-improvement treatments. For example, some early practitioners of applying conformance-improvement preformed-particle-gel treatments involving polyacrylamide-crosslinked gel particles having particle diameters in the range of 1 to 4 mm (0.04 to 0.16 in.), suggested that these treatments were treating conformance problems in matrix-rock reservoirs that had normal permeabilities (<1000 md). However, "rigid" gel particles of such large size cannot be injected into, and propagated through, such reservoir matrix rock. Thus, it was eventually concluded that these preformed particle gels were actually successfully treating high-permeability-anomaly conformance problems, and not high-permeability matrix-rock flow channels. This is an example of how oilfield operators very often tend to grossly underestimate the permeability and/or fluid-flow capacity of the offending conformance-problem flow channels in their oil reservoirs.

For microgels to be readily injectable into, and readily propagatable through, matrix-rock sandstone reservoirs of normal permeabilities (<1000 md), the size of these particles normally must be in the nanometer range. This seems to be a condition satisfied by the technology involving preformed nanosized crosslinked-polymer gel particles that pop/swell-in-size after being placed in an oil reservoir.

However, as briefly mentioned in Section 8.2.1, acrylamide-polymer/aluminum-citrate CDG particles may have particles sizes that are too large for these conformance-improvement-treatment particles to be readily injected into, and propagate through, sandstone reservoir matrix rock of normal permeabilities (<1000 md). If this is true, it must be concluded that either the crosslinking reaction does not take place in the reservoir or another conformance-improvement mechanism is operative for CDG conformance-improvement treatments. One such possible mechanism is that CDG treatments are actually treating intermediate-high-permeability anomalies (e.g., microfractures), and not high-permeability flow channels in sandstone matrix rock of normal permeabilities.

Microfine Cement. Microfine cement is a possible conformance-improvement material that might also best be viewed as the application of a solid-particulate conformance-improvement method. Although it is granted that microfine-cement particles can penetrate further than conventional Portland cement into matrix rock or sands that have permeabilities less than 10 darcies, it is generally accepted that microfine-cement particles cannot normally be placed any significant distance into such reservoir matrix materials (probably not much further than a foot). When it comes to considering the application of microfine cement for application to conformance-problem vertical fractures having normal widths [on the order of 0.5 mm (0.02 in.) or larger], just as with conventional Portland cement for this application, microfine cement often suffers from gravity-settling and particle-bridging problems while being placed, as is often needed, deeply in such conformance-problem fractures.

The application of microfine cement for conformance-improvement use has at times been highly promoted by several oilfield service companies. It can be reasonably argued as to whether the use of conformance-improvement microfine cement should best be discussed as a cement conformance-improvement technology or as a solid-particulate conformance-improvement technology.

Chapter 9

Foams for Conformance Improvement

9.1 Introduction to Foam Use for Conformance Improvement

Foams are another fluid system that can be used for oilfield conformance-improvement purposes (Sydansk 2007; Rossen 1996; Schramm and Wassmuth 1994). As it relates to conformance improvement, foams, when applied within oil reservoirs, have both:

- A viscosity-enhancement component in the foam oil-recovery drive fluid as the foam fluids propagate through reservoir matrix rock
- A permeability-reducing component that can be exploited for conformance improvement

Some early developers of foam flooding anticipated that this technique could become an alternate means to polymer flooding in order to render conformance control during oil-recovery flooding operations. However, the application of foams as viscosity-enhancement drive fluids did not generalize. Foams were especially attractive for use during gas-flooding operations.

Recently, the use of foams as a permeability-reducing conformance-improvement agent for reducing excessive gas production at production wells has been gaining popularity and is beginning to overshadow the use of oilfield foams for viscosity-enhancement of oil-recovery drive fluids during gas oil-recovery flooding operations.

Foams are perceived to have substantial potential for conformance-improvement uses in oil reservoirs. Partly because of this upside potential and partly because foams are unusually complex fluids for conformance improvement in matrix-rock oil reservoirs, foam applications for conformance-improvement have been extensively studied and reported on in the petroleum engineering literature.

On the downside, foams for conformance improvement to date (2011) have not been applied widely in a successful, commercially attractive, and profitable manner in oil fields. There is some possibility, however, with the further development of nanotechnologies, that there might evolve a technological breakthrough relating to conformance-improvement foams and that there might evolve the revitalization of large-scale interest in conformance-improvement foams.

Essentially, all presently (2011) used foams for conformance improvement in oil reservoirs are aqueous-based foams.

Foam application, for gas-flooding mobility-control sweep-improvement purposes and for gas/water permeability-reduction purposes, is considered to be an advanced and nonroutine conformance-improvement technology.

9.2 Nature of Bulk Foams

Foam in bulk form—as in a bubble bath or in the foam head of a glass of beer—is a metastable dispersion of a relatively large volume of gas in a continuous liquid phase that constitutes a relatively small volume of the foam. An alternate definition of bulk foam is an agglomeration of gas bubbles separated from each other by thin liquid films. The gas content of classical foams is quite high (often 60 to 97 vol%). Bulk foams are formed when gas contacts a liquid in the presence of mechanical agitation.

Fig. 9.1 depicts a two-dimensional slice of a generalized bulk-foam system. The thin liquid films separating the foam gas bubbles are defined as foam *lamellae*. For the foam types used for conformance improvement in oil reservoirs, the thin liquid-lamellae structure of the foam is stabilized by the presence of surfactant molecules. The connection of the three lamellae of a gas bubble at a 120° angle is referred to as the *Plateau border*. Foaming agents are often chemical surfactant molecules that are added to (or possibly naturally found in) the foam's liquid phase and are the chemical materials that promote the liquid phase to form metastable foams after being subjected to mechanical agitation.

9.3 Foams in Matrix-Rock Reservoirs

Predicting how foams will exist and function in porous media is not always intuitively obvious on the basis of how foams behave when existing in bulk form (e.g., in a beaker) (Kovseck and Radke 1994).

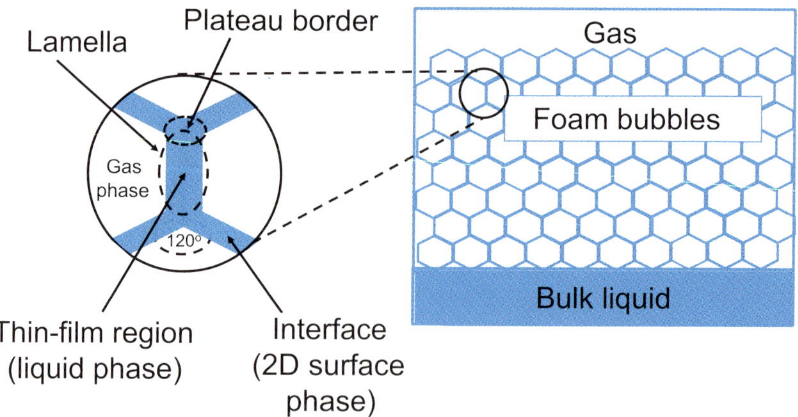

Fig. 9.1—A generalized depiction of a 2D slice of bulk foam. Adapted with permission from Schramm and Wassmuth (1994). Copyright 1994 American Chemical Society.

Capillary processes control the formation and the properties of foams in reservoir-rock porous media. Foams for application as conformance-improvement agents in matrix-rock oil reservoirs are dispersions of gas microbubbles, where the foam's gas bubbles usually have diameters/lengths ranging between 50 and 1000 μm. Foam in porous media exists as individual gas bubbles that are in direct contact with the wetting fluid of the pore walls, where these in-situ gas microbubbles are separated by liquid surfactant-stabilized thin-film lamellae that bridge the pore walls and form a liquid partition on the pore scale between the individual in-situ foam gas bubbles. In matrix-rock reservoirs, foam propagates as bubble trains, where the Plateau border of the foam lamellae is formed at the pore wall and, in many instances, individual foam bubbles can be several pore bodies in length. For static non-flowing foam lamellae in the pore bodies, the average angle between the liquid lamellae and the pore wall is $90°$.

Foams for use in oilfield conformance-improvement operations have surfactants dissolved in the foam's liquid phase in order to stabilize the gas dispersion in the continuous liquid phase. The gas phases of oilfield foams can be either a classical gas or a supercritical gas, such as supercritical/dense CO_2.

9.4 Foams for Mobility Control

Foams can impart a dramatic mobility reduction for gas flow in reservoir matrix rock. Such gas-flow mobility reduction can be up to a factor of several hundred folds. The mobility of the gas phase within the foam as it flows through porous media is greatly reduced, as compared to the mobility of the same gas flowing alone with no foam present. Foam mobility in foam-saturated reservoir matrix rock can be less than the mobility would be for the aqueous phase or the gas phase alone that is used in the foam formulation. For foam saturated matrix reservoir rock, only a small percentage—typically 1 to 15%—of the foam gas saturation actually flows. The stationary portion of the foam gas saturation blocks gas flow in intermediate and small sized flow paths and lowers the effective permeability of the rock to gas flow.

Fig. 9.2 depicts conceptually and schematically foam morphology and flow in porous media—namely, in this case, a sand or glass-bead pack.

In foam-saturated reservoir matrix rock, foam mobility reduction results from a combination of foam-induced permeability reduction, and, on the macro scale, apparent foam-induced viscosity enhancement. Foam-induced mobility reduction is caused by at least two different mechanisms:

- The formation of—or an increase in—the trapped residual-gas saturation
- Increased resistance to flow of the gas phase resulting from the drag of having to propagate foam-lamellae aqueous films through pore bodies, and especially through constricted pore throats

Foam stability and flow in porous media is often strongly influenced by the *limiting capillary pressure, $P_c{}^*$*, above which foam coalescence (destruction) is significant and below which foam coalescence is minimal (Sydansk 2007).

There exist significant questions about the ability to deeply propagate and deeply place foams within matrix rock of oil reservoirs for conformance-improvement mobility-control purposes. One aspect of this concern is the substantial critical pressure gradient that is normally required for initiating and maintaining foam flow in porous

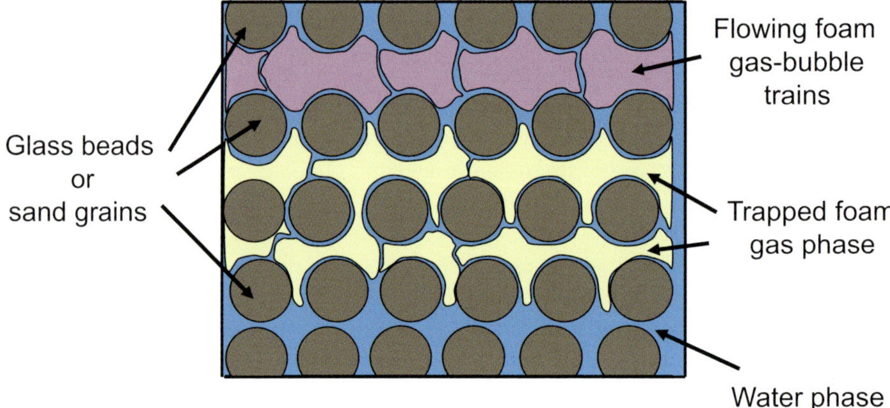

Fig. 9.2—Pore-level schematic of a flowing gas-bubble train and the trapped gas saturation during foam flooding in porous media (flowing gas bubbles are lavender, trapped gas is yellow, the aqueous liquid phase is blue, and solid phase is gray). Adapted with permission from Kovseck and Radke (1994). Copyright 1994 American Chemical Society.

media. This pressing question, derived from the fact that there is a critical pressure gradient required for foam to flow, is as follows. Can foam flow be maintained in the far-wellbore region where pressure gradients are inherently low?

The often negative interaction of crude oil with foam within porous media is another major concern (Schramm 1994). When oil contacts foam, the oil often has a destabilizing effect on the foam. The probability that crude oil will destabilize foams has been a major impediment to the widespread use of foams for oilfield conformance-improvement applications. The destabilizing effects of oil on aqueous foams while flowing through porous media can range from minor to very deleterious—often highly depending on oil properties.

In addition, surfactant adsorption and retention in porous media is a critical aspect to the in-situ performance and the economics of most foam applications for reducing mobility and improving sweep efficiency during gas-flooding operations. Often, the degree of surfactant adsorption and retention can negatively impact the performance of foams and the economics of the application of foams for mobility control during gas flooding. *Retention* is the combination of all other mechanisms, other than adsorption, that consumes surfactant or retards surfactant propagation during foam propagation through reservoir matrix rock. Surfactant adsorption/retention should be one of the key factors to be considered when selecting a surfactant for use in a foam formulation that is to be applied for mobility control.

It should also be noted that the compression cost for injecting preformed foam, or its gas phase separately, into an oil reservoir can be significant and can adversely affect the economics of applying foams for conformance improvement.

9.5 Foams for Gas-Blocking Treatments

Because foams are an effective agent for reducing gas permeability, foams are candidates for use in gas-blocking treatments that are to be placed relatively near-wellbore around production wells. During such application, foams are placed for the purpose of shutting off excessive, deleterious, and competing gas production during oil-recovery operations. Because of the relatively low density of the foam gas-blocking agent, the foam's low effective density provides a driving force for the foam, in concept, to be selectively placed high up in the producing interval where the gas production, especially during gas coning and cusping, is usually flowing into the production wellbore.

The obvious and major challenges, which must be overcome in order to successfully apply foams as a gas-blocking agent surrounding a production well, are to assure that:

- The emplaced blocking foam will have adequate strength to function as an effective gas-blocking agent
- The foam will be stable long enough to make the foam economically attractive for use as a gas-blocking agent.

Advanced Issues Discussion Box

Mechanisms of Foam Formation in Porous Media

A more advanced discussion of the mechanisms of foam formation and destruction in porous media is presented here.

In porous media, foam generation is largely a mechanical process and is less dependent on surfactant formulation and concentration (Apaydin et al. 1998). Foam bubbles are molded and reshaped by the porous media. As bubbles and lamellae are mobilized some distance through a porous medium, perhaps through several pore-bodies and pore-throats, they are destroyed and reformed. Bubble trains immobilize when the local pressure gradient is insufficient to keep them mobilized and other trains convey into motion. The identity of a single bubble or train is not conserved over any large distance. Bubble-trains exist only in a time-averaged sense. The interplay of bubble generation and coalescence determines foam microstructure and, hence, gas mobility in porous media. Three fundamental pore-level generation mechanisms for foam in porous media have been identified as snap-off, lamellae division, and leave-behind (Kovscek and Radke 1994).

Snap-off is a mechanical process that occurs when a bubble enters a pore constriction. The snap-off occurrence is depicted in **Fig. 9.3**, which shows a gas finger that enters a pore constriction initially filled with the wetting liquid (Fig. 9.3a). The pore is considered cornered in cross section with local transverse inscribed radius R_C upon reaching the throat; the interface curvature and corresponding capillary pressure rise to the equilibrium entry value. As the bubble front enters the downstream body, wetting liquid remains in the corners, and the curvature and corresponding local capillary pressure at the bubble front fall with the expansion of the interface. The resulting gradient in capillary pressure initiates a gradient in the liquid pressure directed from the pore-body toward the pore-throat. Liquid is driven along the corners into the pore-throat where it accumulates as a collar (Fig. 9.3b). Collar growth causes the snap-off of the bubble

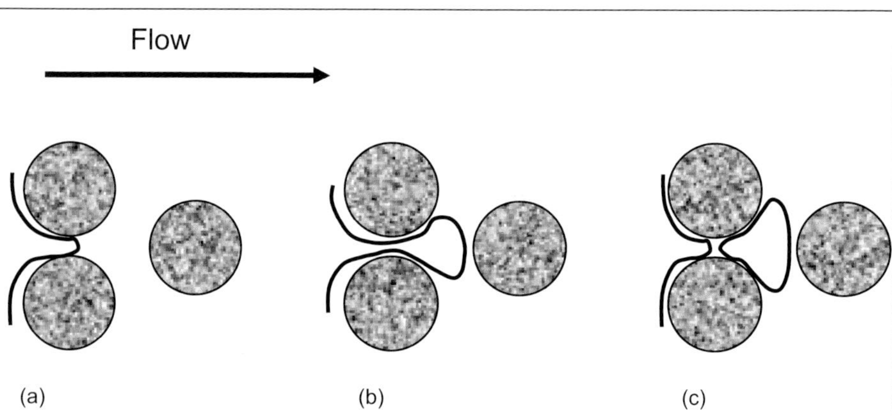

Fig. 9.3—Snap-off mechanism. Adapted with permission from Kovscek and Radke (1994). Copyright 1994 American Chemical Society.

(Fig. 9.3c), if the downstream bubble front has a mean radius of curvature that is larger than about twice RC, and if the constriction is not extremely sharp. On the contrary, if the constriction is too sharp, collar growth is prevented and snap-off will not occur. Snap-off generates bubbles that are approximately the size of pore bodies (Kovscek and Radke 1994). Snap-off bubble formation is dominant in highly heterogeneous pore systems and it is dependent on the degree of permeability contrast (Shirley 1988; Tanzil et al. 2002).

Lamella or bubble division proceeds by subdividing foam bubbles or lamellae as illustrated in **Fig. 9.4.** In this mechanism, mobile foams must pre-exist so that a flowing foam bubble encounters a point where flow branches in two directions (Fig. 9.4a). The interface stretches around the branch point and enters both flow paths. The initial bubble divides into two separate bubbles (Fig. 9.4b) that continue to move downstream. Some of the conditions for lamella division to occur are (1) the bubble size has to be larger than that of the pore body for the foam lamellae to span the pore space and promote division; (2) the surroundings of the branch point should be free of trapped foam bubbles, otherwise, stationary bubbles or lamellae act as flexible pore-walls inhibiting division; and (3) the rate of division events increases as the gas-flow rate increases (Kovscek and Radke 1994).

Leave-behind occurs when two separate gas menisci invade adjacent liquid-filled pore-bodies as shown in **Fig. 9.5a.** A lens is left behind as the two menisci converge downstream. As long as the capillary pressure of the medium is not too high, and the pressure gradient is not too large, a stationary stable lens emerges (Fig. 9.5b). Later, the lens may drain to a thin film. Lenses created by leave-behind are generally oriented parallel to the local direction of flow, and do not make the gas phase discontinuous (Kovscek and Radke 1994).

Lamella created by leave-behind block gas flow channels, hence reducing the relative permeability to the gas phase. The surrounding gas phase, however, remains continuous because no discrete flowing foam bubbles are created.

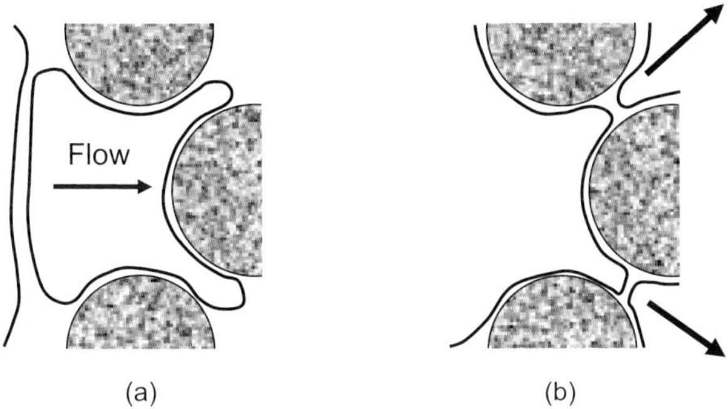

Fig. 9.4—Lamellae division. Adapted with permission from Kovscek and Radke (1994). Copyright 1994 American Chemical Society.

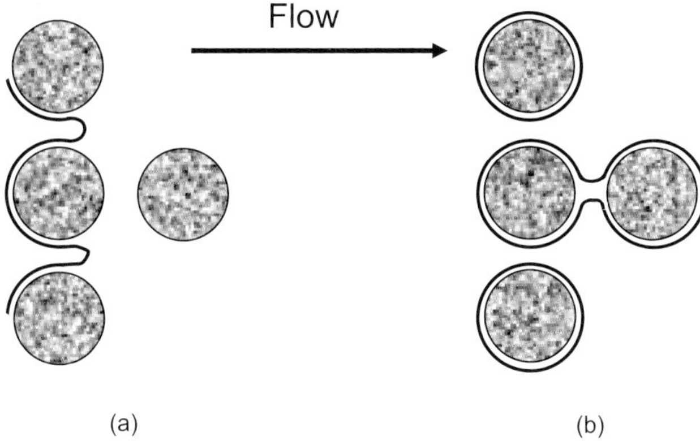

Fig. 9.5—Leave-behind mechanism. Adapted with permission from Kovscek and Radke (1994). Copyright 1994 American Chemical Society.

Unless ruptured by a further increase in the capillary pressure, or mobilized by an increase in the pressure gradient, the leave-behind lamellae remain at their point of generation in the porous medium (Chambers and Radke 1991).

A supplementary mechanism of lamella generation, the bubble-to-bubble lamellae division, which was first identified by Shirley (1988) and was later recognized by other researchers (Romero-Zerón 2004), takes place when a flowing bubble encounters another bubble flowing very slowly (semi-stationary) and perpendicular to the path of the flowing bubble. The flowing bubble compresses the semi-stationary bubble against the pore wall, rupturing the lamellae and producing two bubbles, as illustrated in **Fig. 9.6.** Thus, a long bubble moving in one pore is broken up into new bubbles as a result of colliding with another semi-stationary bubble in intersecting pores (Nguyen et al. 2000). Bubbles formed by this mechanism are several times larger than the pore radius (Shirley 1988).

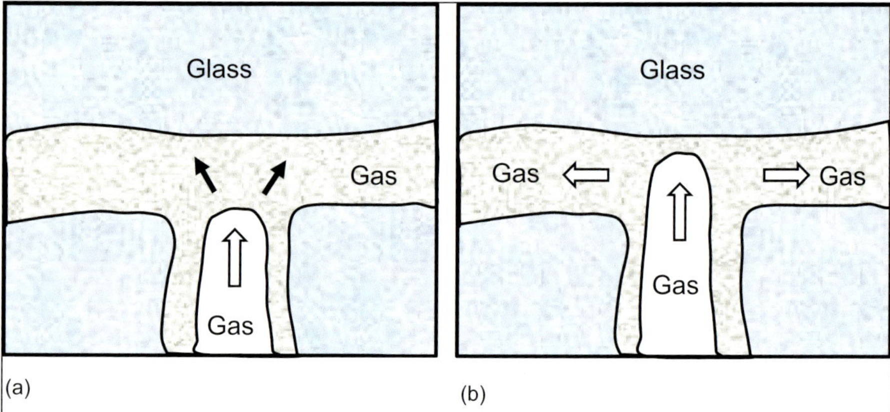

Fig. 9.6—Bubble-to-bubble lamellae division. Adapted with permission from Shirley (1988). Copyright 1988 American Chemical Society.

Mechanisms of Foam Destruction in Porous Media. The main mechanisms of foam coalescence in porous media are *capillary-suction* and *gas diffusion*. Capillary-suction coalescence is briefly reviewed for two specific cases—first, a discussion of the coalescence of the static trapped lamellae is presented, followed by the coalescence behavior of flowing foams. Gas diffusion is discussed last.

Coalescence of Static Trapped Lamellae. The thickening of static trapped lamellae in porous media is represented by the Young-Laplace equation as follows.

$$P_c = 2\sigma C_m + \Pi(h_f) \quad\dots\dots\dots\dots\dots\dots\dots\dots\dots\dots\dots\dots\dots \quad (9.1)$$

In Eq. 9.1, P_c is local-capillary pressure, C_m is the mean interfacial curvature of the thin film, σ is the bulk surface tension, and Π is the film disjoining pressure, which is a function of film thickness, h_f. At the limit of thick films the disjoining pressure approaches zero, and the classic Young-Laplace equation is recovered (Kovscek and Radke 1994). For the static trapped lamellae, the film reaches an equilibrium thickness set by the local capillary pressure and the film curvature. The capillary pressure in turn depends on the wetting-liquid saturation; the film curvature depends on the particular location within the pore structure dictated by an approximately 90° contact angle with the pore wall. As the capillary pressure in the porous medium rises during drainage, the film thickness decreases until a maximum film disjoining pressure, Π_{max}, is reached. The film thickness matching Π_{max} is designated as h_{fmax}. At higher imposed capillary pressures, a rupture disjoining pressure, Π_{rup}, is reached, where the film eventually ruptures. Thus, static lamellae are stable to small disturbances until a critical capillary pressure for rupture is attained; then coalescence is catastrophic.

Coalescence of Flowing Foams. According to Kovscek and Radke (1994), during steady-state foam flow, drastic foam coarsening exists at a specific capillary pressure, called the limiting capillary pressure, P_c^*. Thus, above P_c^*, coalescence of flowing lamellae is significant and below P_c^*, it is minimal. Limiting capillary pressure is a function of gas flow rate, absolute permeability, and surfactant formulation. The *hydrodynamic stability theory* links the limiting capillary pressure and the critical disjoining pressure for lamellae rupture. **Fig. 9.7** illustrates a lamella flowing through a periodically constricted tube at successive times t_1, t_2, and t_3. As the lamella moves from left to right, it is squeezed upon entering the constriction at time t_2. At this location, film thickness increases to conserve liquid mass, and the disjoining pressure, Π, is correspondingly low. When the lamella moves out of the pore constriction, it is stretched upon expansion into the downstream body at time t_3. The film thins to conserve mass, and the disjoining pressure is now high. Here, the film fills in an attempt to equalize P_c and Π, again driven by the pressure difference $(P_c-\Pi)$. Thus, the thickness of the transporting lamella oscillates around the equilibrium thickness established in the stationary lamella in a sequence of squeezing-stretching and draining-filling events. A flowing lamella breaks instantaneously (at time t_3) whenever the film thickness diminishes to $h_{f\mathrm{max}}$, corresponding to Π_{max}. Thus, if the film is stretched too rapidly for healing surfactant solution to flow into the film and stabilize it, rupture results. The capillary pressure at which the flowing lamella ruptures is the limiting capillary pressure, P_c^*. Limiting capillary pressure, P_c^*, increases rapidly with larger pore-body to pore-throat aspect ratio and increases slowly with higher gas flow rate. In addition, low wetting-phase saturation, which corresponds to high capillary pressure, promotes the rapid stretching of lamellae to cause rupture, even at small aspect ratios because there is not enough wetting phase (surfactant solution) available to stabilize the lamellae. Thus, capillary-suction coalescence depends greatly on surfactant formulation. If surfactant solution exhibits high rupture disjoining pressures, then strong foams in porous media with high resistance are expected.

Gas Diffusion. Diffusion is the dominant mechanism that leads to the coalescence of stationary foams under conditions in which all lamellae are exposed to capillary suction pressures significantly below the limiting capillary pressure (Kovscek and Radke 1994). According to Cohen et al. (1996), foam injected into a porous medium behaves quite differently than in bulk form, due to the geometric constraints at pore walls. In a water-wet porous medium, the projections of

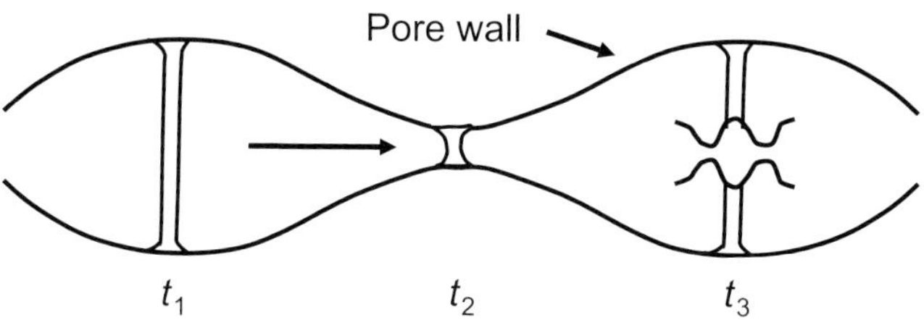

Fig. 9.7—Lamella flowing through a periodically constricted tube at successive times t_1, t_2, and t_3. Adapted with permission from Kovscek and Radke (1994). Copyright 1994 American Chemical Society.

the lamellae intersect the walls at a 90° angle. Therefore, the curvature of a given lamellae is completely determined by its location within the pore. It is no longer necessarily true that small bubbles are at higher pressure than large bubbles. Thus, gas diffuses through high pressure lamellae to low pressure lamellae. As a result, lamellae rearrange as the system strives to achieve a state of minimum free energy, at which point the pressures become uniform. Lamellae, which are assumed to be infinitely stable, move by sliding along the wetting films, which coat the surface of the porous medium. This process proceeds very slowly allowing the liquid phase film to rearrange so that lamellae remain at nearly constant thickness during the process. At equilibrium position, the lamellae all rest in the pore throats, where they have no curvature and thus sustain no pressure drop. This configuration is thermodynamically stable and gas diffusion stops. In contrast to bulk foam, which always collapses ultimately, the equilibrium configuration of foam in a porous medium remains indefinitely in the absence of external disturbances. The time to reach equilibrium is a quadratic function of the characteristic pore size and a linear function of pressure. At high pressure, the foam takes a much longer time to decay. For thicker lamellae the rate of diffusion will decrease. Hanssen and Dalland (1994) observed very long persistence times for gas-blocking foams, indicating that characteristic diffusion times for these types of foams could be on the order of weeks or months.

9.6 Injecting Foam in the Field

Injection of conformance-improvement foams into a reservoir is usually applied in one of three different modes:

- Sequential injection
- Co-injection
- Preformed on the surface prior to injection

Due to the substantial effective viscosities of foams and the associated relatively poor injectivity of preformed foams, early applications of conformance-improvement foams tended to involve the sequential-injection or co-injection mode. The effective viscosity of preformed foam during injection is normally greater that the viscosity of either the foaming aqueous solution alone or the gas phase of the foam. Thus, in order to attain favorable injectivity for the "foam" to be placed in a reservoir, early practitioners of conformance-improvement foam flooding/treatments often chose to either sequentially inject or co-inject the aqueous foaming solution and the gas. In addition, when conducting foam conformance-improvement operations, the use of the sequential-injection and co-injection mode is often substantially simpler than the use of preformed-foam (from an operational standpoint) to implement in the field setting. Sequential injection also largely avoids tubular corrosion problems if the gas and foaming-solution together form a corrosive mixture—such as found in CO_2 foams.

The concept is, and there is laboratory evidence suggesting, that during the sequential or co-injection mode, foam will form in situ in the reservoir matrix rock. Supporting this contention is the expectation that the low-viscosity and high-mobility gas will tend to finger into the aqueous foaming solution and thus presumably generate the foam in situ. However, there are two significant countering concerns. First, as the gas begins to finger into the aqueous solution and initially begins to form foam in situ, the newly formed foam, which is an effective permeability-reducing agent to gas flow, will effectively and substantially reduce subsequent gas fingering. The newly formed foam will divert subsequent gas flow away from the remaining aqueous foaming solution residing just ahead of where the initial foam was formed. Thus, this phenomenon renders ineffective and inefficient the utilization of the injected foam chemicals and fluids in terms of effective generation and penetration of foam deep in the reservoir formation for the intended conformance-improvement purpose.

Second, under intermediate and far wellbore differential-pressure conditions, there may not be enough mechanical energy and/or differential pressure present to generate foam in situ when employing commonly used oilfield foaming solutions. This is especially of concern for steam, nitrogen, and natural-gas foams.

In the authors' view, unless compelling arguments for a specific application can be made to the contrary, foams for most applications in the *near and intermediate wellbore region* should be preformed at the surface prior to injection.

The foam sequential process, alternatively known as the foam water-alternating-gas (FWAG) process, of injecting sequentially and repeatedly alternating slugs of gas and aqueous foaming solution is often favored when using CO_2 foam for far-wellbore mobility-control purposes during CO_2 flooding. This is because CO_2 dissolved in the aqueous surfactant solution of preformed CO_2 foams form carbonic acid that is corrosive to steel injection tubulars. Because of the low surface tension between CO_2 and the aqueous foaming solution, foam generation in situ in reservoir matrix rock is much more feasible (than steam, nitrogen, or natural-gas foams) at the realistic field pressure gradients that occur throughout the reservoir.

9.7 Foam Field Application for Conformance Improvement

The successful application of foams for conformance-improvement use is limited to placement in *matrix-rock* oil reservoirs. To present (2011), there have been three major field applications of conformance-control foams:

- Mobility-control agent during steamflooding (Hirasaki 1989)
- Mobility-control agent during CO_2 flooding (Heller 1994)
- Gas permeability-reducing agent when placed around production wells—especially in conjunction with miscible-gas-flooding projects (Sydansk 2007)

The low effective density of most mobility-control foams that are used during a gas flood such as a steam or CO_2 flood provides a driving force for the foam to be (desirably) placed high up in the reservoir vertical interval where offending gas override is often occurring. This is where the foam will be most effective at countering the negative impact on oil-recovery sweep efficiency of the gas flooding override. Similar foam-density arguments apply when using conformance-improvement foams to block gas coning or cusping.

9.8 Advantages and Disadvantages of Foam Use in Conformance Control

A number of advantages and disadvantages exist regarding the use of foams for improving conformance during oil-recovery operations.

Advantages:

- Foams are an exceptionally effective technology for reducing gas mobility during gas flooding.
- Foams can be an effective gas-blocking agent.
- Foams are a conformance-improvement material that has the tendency in certain instances to reduce permeability and mobility to a greater degree than other conformance-improvement technologies (e.g., gels, polymer) in higher-permeability matrix reservoir rock.
- Foams are shear-thinning fluids—resulting in relatively good injectivity and resulting in more effective mobility control in the far-wellbore region where such mobility control is most needed.
- Foams possess low effective density that can often be exploited to help selectively place them high up vertically in a reservoir so as to impede problematic gas flow where it is most likely to occur.
- Foams are considered, in general, to be an environmentally friendly material for use in conformance-improvement operations.

Disadvantages:

- Foams are a relatively complex technology, both chemically and operationally, making them difficult to successfully apply in the oilfield.
- Oil tends to destabilize and deactivate many conformance-improvement foams.
- For many mobility-control foams (e.g., steam and natural gas foams), there exists a difficulty and/or inability to propagate the foams in the intermediate to far wellbore environment under the differential-pressure conditions encountered in most reservoirs.
- For mobility-control foams, surfactant adsorption/retention has a substantial negative impact on the performance and the economics of foam-flooding operations.
- For fluid-blocking (e.g., gas-blocking) foams used in production-well treatments, the emplaced foams have limited strength under high-differential-pressure conditions.
- Fluid-blocking (e.g., gas-blocking) foams used in production-well treatments are limited by the inherent lack of long-term foam stability and the associated lack of long-term treatment effectiveness.
- The relatively high viscosity and associated relatively poor injectivity of preformed foams limit the application of this otherwise favored foam-injection mode.

- The limited and sometimes poor ability to effectively form foam in situ (in matrix reservoir rock) during the co-injection or sequential injection of the foam's gas and aqueous-liquid phases reduces the effectiveness and the efficiency of the co-injection and sequential-injection modes for foam formation and placement in a reservoir.
- High compression costs associated with injecting preformed foams and/or injecting the gas and liquid phases separately during the sequential or co-injection modes, can detract from the economics of applying foams for conformance-improvement purposes.

9.9 Advanced Foams for Conformance Improvement

Polymer-enhanced foams and foamed gels are advanced forms of foam that have been proposed and studied for conformance-improvement use in oil reservoirs.

The concept of polymer-enhanced foams is that the addition of a water-soluble high-molecular-weight viscosity-enhancing polymer (e.g., an acrylamide polymer) to the aqueous phase will improve the stability of the foam, increase the effective viscosity of the foam, and improve the overall oilfield performance of the foam. There have been several engineering literature papers both supporting and discrediting the concept of using conformance-improvement polymer-enhanced foams (Huh and Rossen 2008; Sydansk 2007; Shen et al. 2006).

The concept of conformance-improvement foamed gels (Sydansk 2007) is as follows. The addition of significant volumes of gas to oilfield conformance-improvement polymer gels would greatly reduce the unit volume cost of permeability-reducing gels. Also, the reduced density of foamed gels would allow them to be preferentially placed high up in open natural fractures to block conformance-problem gas flow high up in the fractures of the type that often occur when CO_2 flooding is applied to naturally fractured reservoirs. CO_2 IOR/EOR flooding in the United States has historically been most widely applied to naturally fractured reservoirs.

There are two challenges affecting the wide-spread application of the conformance-improvement use of foamed gels. First, foamed-gel conformance-improvement treatments are more complex to formulate and apply than the counterpart straight polymer-gel treatments. Second, because of the relatively low density of foamed gels (and consequential reduction of foamed-gels' hydrostatic contribution to the injection pressure), it takes substantially more horse power to inject foamed gels as compared to analogous straight aqueous polymer gels.

9.10 Status and Prospects for Foam Conformance Control

Foam application for conformance improvement in oil reservoirs is presently not being widely implemented. Over recent years, the focus of foam application in oil reservoirs has shifted from foam use as a mobility-control agent during oil-recovery flooding operations (especially miscible gas flooding) to foam application as permeability-reducing and fluid-flow blocking agents for blocking gas production at production wells.

The current outlook for using foams extensively and routinely for improving conformance during oil-recovery flooding operations is not as encouraging as it was during the past three and a half decades. Petroleum industry R&D in the area of conformance-improvement foams has been diminishing. The original interest in foams as mobility control agents for application during oil-recovery flooding operations has faded somewhat. Because of the realization that foam fluid-flow blocking treatments are somewhat operationally and chemically complex to apply and the realization that there exist alternate, highly competitive, and more effective/stronger fluid-flow blocking agents for use in conformance-improvement treatments (e.g., polymer gels), interest in the use of conventional foams as fluid-flow blocking agents has also started to fade somewhat.

Chapter 10

Selected Field-Application Examples of Conformance-Improvement Techniques

This chapter will briefly review and discuss six selected successful field applications of conformance-improvement techniques. All of these field applications have one thing in common: The conformance-improvement techniques that were selected and applied had an exceptionally good fit between the reservoir conformance problem to be treated and the selected conformance-improvement technique.

10.1 Daqing Oil Field Polymer Flood

Reservoir properties in the Daqing oil field in China favored polymer flooding with HPAM polymer (Wang et al. 2008b). The sandstone reservoir has an average permeability slightly greater than 1 darcy. The reservoir has a mild temperature of 113°F (45°C). The total salinity of the reservoir formation water varies between 3,000 and 7,000 ppm. Oil viscosity at reservoir temperature averages 9 cp.

Conformance-improvement polymer waterflooding using HPAM (hydrolyzed polyacrylamide) polymer in the Daqing oil field was reported in 2002 to have recovered cumulatively over 300 million bbl of incremental oil production (Demin et al. 2002a). The Daqing polymer waterflooding project was reported to have produced 70 million bbl/yr of incremental oil production in 2001 at a rate of 310% greater than that expected for waterflooding without a polymer. Also at that time, the ultimate incremental oil production attributed to the Daqing polymer waterflood was projected to be in the range of 12 to 15% of the OOIP. The cost of the incremental oil production from the polymer waterflooding project was stated to be USD 6.60/bbl.

Additionally in 2002, it was reported that about 76% of the total Daqing oil production was derived from polymer flooding IOR (Demin et al. 2002b). Similarly, it is claimed that the large-scale implementation of polymer flooding at the Daqing oil field had more than doubled the recoverable reserves of the field from approximately 7 billion barrels to approximately 16 billion barrels of oil.

Wang et al. (2008b) reported that in 2007, 23% of the *total oil production* at the Daqing field was attributed to polymer flooding and that the polymer flooding should boost the ultimate oil recovery in the Daqing field to greater than 50% of the OOIP. In 2007, oil production resulting from polymer flooding at Daqing was reported to be 73 million bbl/yr, and this incremental-oil-production rate has been maintained at roughly this level for the previous six years.

10.2 Gel Injection-Well Fracture-Problem Treatments for Improving Sweep

The chromium(III)-carboxylate/acrylamide-polymer (CC/AP) gel conformance-improvement technology was conceived and developed in the mid 1980s. This conformance-improvement gel technology was first applied and demonstrated in the Permian basin of west Texas and in the Big Horn basin of Wyoming with the majority of these early gel treatments being applied in the Big Horn basin (Sydansk 2007; Sydansk and Southwell 2000, Sydansk and Moore 1990).

The first 17 CC/AP gel conformance-improvement treatments, which involved the use of chromic triacetate as the crosslinking agent and were applied to fractured injection wells in the Big Horn basin for the purpose of improving sweep efficiency during oil-recovery flooding operations, are reviewed in this section. These gel treatments were applied between 1985 and 1988. The gel treatments were applied to two carbonate formations (Embar and Madison formations) and to a single sandstone formation (Tensleep formation). The treatments were applied in three different fields. The three reservoirs are characterized as having fracture networks of intermediate density, where the fracture networks exhibit directional trends.

These 17 CC/AP gel sweep-improvement treatments ultimately recovered 3,650,000 bbl of incremental oil or, on the average, 215,000 bbl of incremental oil per gel treatment. The production cost of the incremental oil, on the average, was reported to be USD 0.21/bbl. During these CC/AP gel treatments, 13.6 bbl of incremental oil were, on the average, produced for every pound of polymer employed within the gels. Stated another way, about 4,000 tons of oil were recovered per ton of polymer used within the gels.

Of these 17 CC/AP gel injection-well treatments, 11 were applied to carbonate reservoirs and 6 to a sandstone reservoir. The average incremental oil recovery and the production cost of the incremental oil for the gel treatments applied to the carbonate reservoirs were, respectively, 261,000 bbl and USD 0.18/bbl, vs. 129,000 bbl and USD 0.33/bbl for the gel treatments applied to the sandstone reservoir. It should be noted that the fracture conformance problems in the carbonate reservoirs were believed to be more amenable than those in the sandstone reservoir to positively responding to the gel sweep-improvement treatments.

Fig. 10.1 shows the production response to one of the first half-dozen CC/AP gel treatments ever conducted. The figure shows the combined production response of the four direct offsetting production wells to the gel-treated injection well. This 20,000-bbl gel treatment was applied for oil-recovery-drive-fluid sweep-improvement purposes to the naturally fractured Embar carbonate formation surrounding Well O-7 of the South Oregon basin (SOB) field in the Big Horn basin. The wide variations in WOR and oil production rate shown in Fig. 10.1 prior to the gel treatment are quite common in many of the well patterns of this highly naturally fractured reservoir.

The sweep-improvement CC/AP gel treatment that was applied to the fractured injection Well O-7 in Fig. 10.1 ultimately generated 530,000 bbl of incremental oil

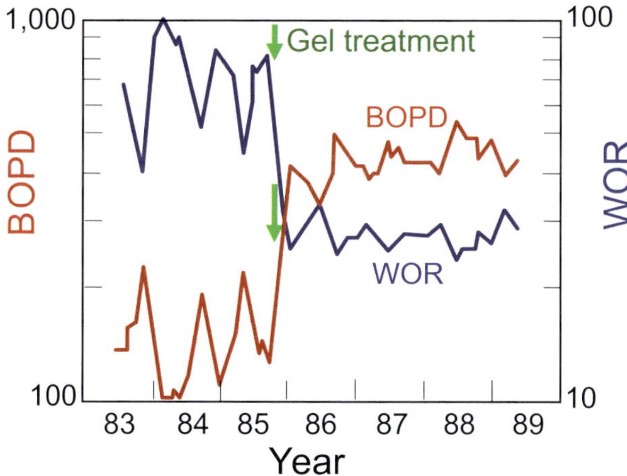

Fig. 10.1—Production response to the CC/AP gel treatment applied to injection-well O-7 in the SOB field. Adapted from Sydansk and Moore (1990).

production. The cost of this field-demonstration gel treatment was USD 88,000. The gel treatment cost included all gel chemical and pumping costs, plus well-workover costs and field-testing costs. Because of the substantial costs associated with the field-testing aspects of this treatment, the gel-treatment cost in this case was exceptionally high. Despite the exceptionally high cost of the treatment, the reserve development cost for the incremental oil production was only USD 0.17/bbl. Note that as shown in Fig. 10.1, the CC/AP gel treatment generated a large incremental oil-production rate that persisted for years.

10.3 Production-Well Fracture Problem—Gel Treatments for Water Shutoff

At the present time (2011), almost 1,000 CC/AP gel water-shutoff treatments have been applied to vertical production wells of the fractured-carbonate Arbuckle formation within the state of Kansas. The vast majority of these gel water-shutoff treatments have been conducted since the year 2000. Typically, 1,000 bbl (or more) of gel were injected during these conformance-improvement water-shutoff CC/AP gel treatments.

The excess water production from the treated Kansas Arbuckle wells is believed to be water coning up to the wellbore through natural fractures from a strong aquifer underlying the Arbuckle formation. All Arbuckle fields are under primary production. Kansas Arbuckle wells, for the most part, are very old and are marginal production wells—often approaching their economic limit. In 2003, over 90% of the Kansas Arbuckle wells produced less than 5 BOPD at water cuts ranging from 99.0 to 99.5%. Therefore, the CC/AP gel water-shutoff treatments applied to the Arbuckle formation aimed to reduce water production, to increase oil recovery, and to prolong the life of some of the wells.

In 2008, Willhite and Pancake analyzed several of these Arbuckle CC/AP gel water-shutoff treatments in detail and reported that the objectives of the treatments were achieved. For instance, water-production rate in the analyzed wells was successfully reduced by 53 to 90%. Incremental oil production was obtained and when the CC/AP gel water-shutoff treatments were applied to wells completed openhole, the amount of incremental oil production increased with the increasing volume of gel treatment fluid injected (1,500 to 4,000 bbl of gel). Of substantial interest is the fact that the oil productivity index increased following the application of these Arbuckle CC/AP gel water-shutoff treatments.

Likewise, Reynolds (2003) reviewed approximately 300 Arbuckle CC/AP gel water-shutoff treatments that had been conducted since the year 2001 by over 30 different operators. For these gel treatments, pre- and post-treatment production data were available for analysis for 37 of the treated wells. The analysis of the production data for these 37 gel treatments was upscaled to project the average production of the 300 completed gel water-shutoff treatments. The results of this analysis show that the 300 CC/AP gel treatments have shut off 110 million bbl of excessive, unnecessary, unwanted, costly, directly competing, and environmentally unfriendly water production. Additionally, Reynolds (2003) stated that the production costs of the Arbuckle CC/AP gel water-shutoff treatments were in the range of USD 2 to 5 per barrel and that many of these gel water-shutoff treatments paid out in 3 to 7 months (based on incremental oil production).

The success of the CC/AP gel water-shutoff treatments applied to the Arbuckle formation has been further confirmed by other authors. For instance, in 2005, Portwood reported that the success rate for these gel water-shutoff treatments exceeded 95%. Less than 1% of these treatments failed to reduce the water production rate and/or increase the oil-production rate. Portwood (2005) also reported that the average volume of gel injected was 2,300 bbl, the average gel-treatment cost was USD 23,400, and the average cost of the gel injected was USD 10.13/bbl. Another economic analysis conducted by Alhajeri et al. (2006) on a set of 59 of these Arbuckle CC/AP gel water-shutoff treatments indicated an 89% of economic success rate using an oil price of USD 22/bbl. The authors pointed out that the Arbuckle wells provide a good example of how polymer-gel conformance-improvement treatments can extend the life of high-water-cut wells in a cost-effective fashion.

An issue with nearly all successful Kansas Arbuckle gel water-shutoff treatments is that, although they are successful and the operators continue applying them, the high incremental oil-producing rate is normally only maintained for a number of months, or possibly up to a couple of years, after treatment application. Some of the reasons that explain the performance of these Arbuckle CC/AP water-shutoff treatments are as follows.

First, when producing from a well at substantially over the critical coning rate, it is well known that although the chemical blocking agents (CC/AP gels) have been successfully placed in fractures to prevent water coning up through the fractures, eventually the water will cone around the emplaced water-shutoff material. Second, the performance of the CC/AP water-shutoff treatments, including maintaining high incremental oil-producing rates, depends on gel-treatment volume. The excess Arbuckle water production problem is caused by water coning up through fractures. For this specific type of excessive water-production problem it is generally accepted that the larger the volume of the gel treatment injected, the longer lasting will be the treatment's benefits. Therefore, it is possible that the volume of the water-shutoff treatments applied to some of the Arbuckle formation production wells was underestimated.

10.4 Gel Conformance-Improvement Treatments Applied During CO_2 Flooding

Conformance-improvement CC/AP gel treatments have been successfully applied at the Wertz field CO_2 tertiary water-alternating-gas (WAG) flooding project in the Wind River basin of Wyoming. Ten injection-well CC/AP gel treatments were applied to a 165°F naturally fractured Tensleep sandstone reservoir from 1991 through 1993 (Borling 1994). The following benefits were derived from these gel treatments during the Wertz CO_2 flooding project:

- Incremental oil recoveries of up to 140,000 bbl per well pattern
- Increased oil production rates by 100 to 300 BOPD per well pattern
- Extended the economic lives of marginal well patterns by nearly 2 years
- Reduced GORs and WORs
- Reduced gas and water cycling
- Increased gas and water breakthrough times
- Improved water and gas injection profiles
- Reduced operating expenses
- Contributed to the field-wide decline-rate reduction in 1992 from 24 to 9%
- Effective gel treatment performance where conventional oilfield foams had failed
- Rapid payout times of often less than three months
- Recovered substantial oil reserves that would not have been otherwise recovered

Separately, 44 injection-well CC/AP gel treatments were applied during 1994 through 1997 at the large CO_2 miscible WAG flooding project of the Rangely Weber Sand Unit (RWSU) that is located in northwestern Colorado (Hild and Wackowski 1999). At the time of these gel treatments, the RWSU was the largest field in the Rocky Mountain region in terms of daily and cumulative oil production. The economic rate of return for these large-volume (~10,000 bbl) injector gel treatments was 365%. The economic success rate of these treatments was 80%.

10.5 Field Demonstration of Steam Diversion with Foam

A foam-steam diversion application was conducted in a heavy oil reservoir at the South Belridge field located on the western side of the San Joaquin Valley, approximately 45 miles west of Bakersfield, California. As reported by Djabbarah et al. (1990), steamflooding began in this field in an inverted ten-acre nine spots in the northwest areas of the field in 1972 and gradually expanded to 16 patterns by 1979. During this period, the steamflooding operation demonstrated inefficiencies because of heat loss to the overburden (steam override) and because of reduced volumetric sweep efficiency that resulted in a significant amount of unrecovered oil. A foam-steam diversion pilot was initiated in 1987 in two of these patterns (two contiguous ten-acre inverted nine-spot patterns) with the goal of improving vertical and areal sweep efficiency.

Foaming was achieved by co-injecting steam, nitrogen, and surfactant. The surfactant used was a linear toluene sulfonate having physicochemical stability and the ability of generating foam at high flow resistance. A two-phase injection strategy was used to generate and propagate foam. The first phase involved the continuous injection of

1% active surfactant, 25 scfm of nitrogen, and steam for 1 to 2 weeks. The second phase was planned as a 24-hour cycle consisting of 12-on, 12-off 0.5% active surfactant and the continuous injection of 20 scfm of nitrogen with steam. Foam injection began in July 1987 and was terminated in July 1988. During this period, some 179,000 lbm of active surfactant, 15 MMscf of nitrogen, and 443,000 bbl of steam, cold water equivalent basis (CWEB) were injected. This was equivalent to 0.011% PV of surfactant and 6.9% PV of nitrogen (Djabbarah et al. 1990).

The field test results demonstrated that foam formation and propagation effectively diverted steam and improved sweep in the patterns under evaluation. After foam injection, higher temperature occurred broadly over the patterns and evenly in the vertical profile. On the basis of temperature logs run prior to and after foam injection, it was evident that the steam-zone thickness doubled in size, and sweep efficiency increased in the thick zone where steam vertical override was believed most prevalent prior to foam injection. The oil production response clearly demonstrated the feasibility of steam flow diversion by foam in this complex reservoir sands. Incremental oil in the order of 183,000 bbl of oil was produced during the period of foam injection as shown in **Fig. 10.2** (Djabbarah et al. 1990).

10.6 Application of Foam Gel to Improve Conformance During CO$_2$ Flooding

In 1986, a miscible CO$_2$ flood was initiated in the RWSU, located in Rio Blanco County, Colorado, USA (Friedmann et al. 1999). RWSU is the largest field in the

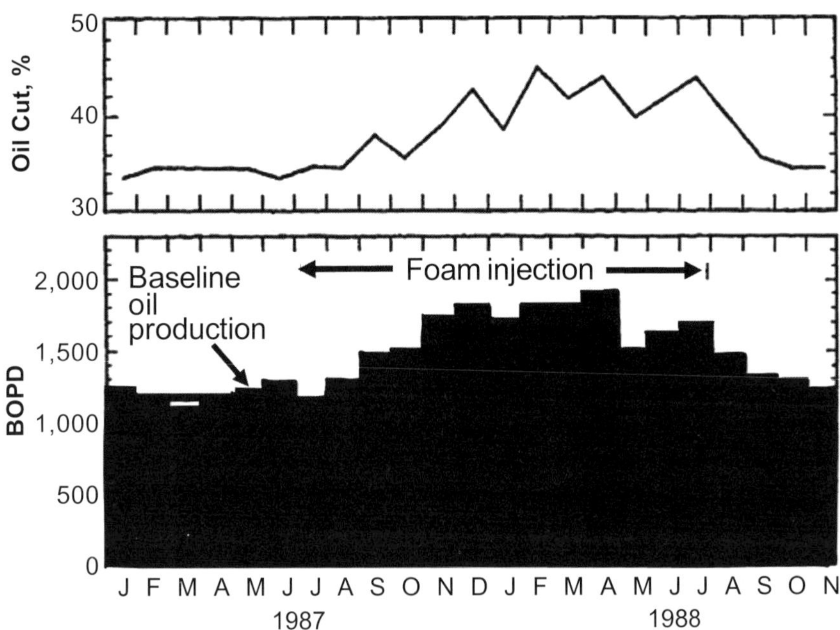

Fig. 10.2—Oil production response to the foam-steam diversion field trial (Djabbarah et al. 1990).

Rocky Mountain region in terms of daily and cumulative oil production. At the time of the CO_2 flooding, the majority of the CO_2 injector wells were under the WAG process. CO_2 was injected in WAG cycles of approximately 2 weeks.

Serious conformance problems were identified in the RWSU causing poor vertical and sweep efficiency attributed to hydraulic fractures (almost all the injectors had been hydraulically fractured), natural fracture networks, reservoir heterogeneity, and injection/production imbalance. The main problem in the RWSU was poor CO_2 conformance causing rapid breakthrough or cycling to the producers via the natural fracture network, resulting in high operating expenses and low oil recovery. Several wells were treated by placing a MARCIT™ gel into the fracture network. The success rate of these treatments was estimated to be over 80%. However, gel treatment volumes were limited to about 15,000 bbl for economic reasons. Although these treatments improved the local sweep of the reservoir and increased recovery, larger treatment volumes were needed in parts of the field where CO_2 breakthrough at offset producers was particularly severe (Friedmann et al. 1999). The approach taken at the RWSU to overcome the treatment volume limitation was the implementation of large-volume foam-gel treatments, which can provide a cost-effective method to achieve in-depth conformance improvement in fractured reservoirs.

During the period of November 1996 to November 1997, three large-volume CO_2-foam-gel treatments were implemented at the RWSU aiming to improve volumetric conformance by reducing excessive CO_2 breakthrough through fractures and to increase oil recovery from the associated producers (Friedmann et al. 1999; Hughes et al. 1999). It is important to emphasize that the field implementation of these CO_2-foam-gel treatments was supported by a systematic laboratory R&D of the gelled-foam system, which was thoroughly tested in the laboratory at reservoir conditions. Treatment I was performed in a relatively large pattern with well-documented rapid CO_2 communication trends in three directions. Treatments II and III were applied in the diagonal line drive region of the field and both patterns had well documented rapid CO_2 breakthrough to only one producer in their respective areas. All the well patterns chosen for application of treatments I, II, and III contained significant bypassed reserves due to poor CO_2 conformance and premature tapering of the WAG ratio. The volume of foam-gel treatment injected ranged from 36,400 bbl (Treatment I) to 44,700 bbl (Treatment III) (Hughes et al. 1999).

The metrics used to evaluate the impact of the gelled-foam treatments on the volumetric sweep throughout the patterns were as follows:

- CO_2 and brine injection profile surveys
- Pressure fall-off tests
- CO_2 breakthrough times at offset producers
- Oil, CO_2, and water production in offset producers

The patterns' production responses indicated that each of the treatments induced a stabilization in the pattern oil rate, which for Treatments I and II was accompanied by a decrease in the pattern gas rate. The first positive oil-rate response given by each of the treatments was observed 6–8 months after foam-gel treatment. **Fig. 10.3** shows the pre- and post-treatment pattern oil production history for Treatment I (Hughes et al. 1999).

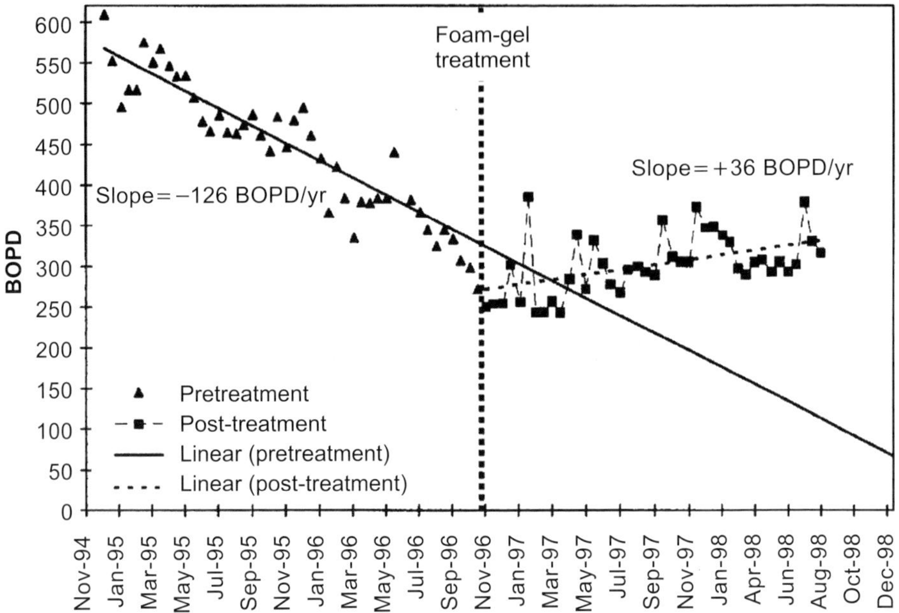

Fig. 10.3—Pre- and post-treatment pattern oil production history for Treatment I (Hughes et al. 1999).

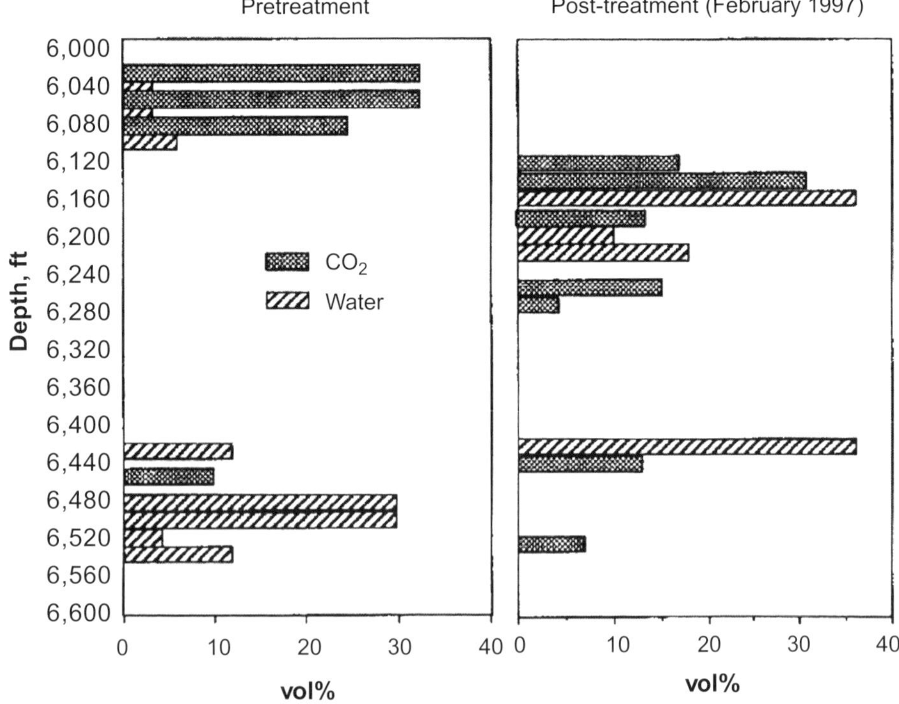

Fig. 10.4—Pre- and post-treatment injection profile surveys performed in L.N. Hagood A9X during Treatment I (Hughes et al. 1999).

Hughes et al. (1999) described the decrease in oil rate by a linear regression with a slope of -126 BOPD/yr prior to Treatment I. During the period December 1994 to May 1996, the WAG-tapering process was evident by the significant decrease in CO_2 injection rate from 5,700 to 3,400 Mscf/D in response to the declining pattern oil rate and CO_2 cycling. Post-treatment, the increase in oil rate was described by a linear regression with a slope of $+36$ BOPD/yr. Thus, the CO_2 foam-gel (in Treatment I) reversed the oil production decline. By the end of August 1998 the incremental oil production attributed to the gelled-foam treatment was approximately 40,000 bbl over a 2-year period (Hughes et al. 1999).

Both CO_2 and brine injection profiles indicated diversion into new sands. **Fig. 10.4** shows the pre- and post-treatment injection profile surveys performed in the L.N. Hagood A9X well during Treatment I. The injection profile surveys in Fig. 10.4 indicate that CO_2 and water entry were diverted from the upper sand zones to the intermediate sand zones, thereby improving vertical sweep efficiency (Hughes et al. 1999).

Pressure fall-off tests indicated a change from linear to radial flow near the injector. CO_2 breakthrough times were increased in some offset producers. For a given treatment volume, the cost of foam-gel treatment at RWSU was 40–50% below the average cost of polymer-gel treatments. As the foam is injected at a higher rate, the total pump time required for a 40,000 bbl foam-gel treatment is similar to the total pump time required for a 20,000 bbl polymer-gel treatment. The economic evaluation of the foam-gel approach at the RWSU by August 1998 indicated that CO_2 foam-gel was a cost-effective method to achieve in-depth conformance improvement in fractured reservoirs (Friedmann et al. 1999; Hughes et al. 1999).

Chapter 11

Conformance During Enhanced Oil Recovery

11.1 Background Information

This chapter addresses the issue that conformance and conformance problems are often the major pitfall to the successful implementation and application of enhanced-oil-recovery (EOR) and improved-oil-recovery (IOR) flooding operations.

Although the industry interest in EOR flooding had fallen out of favor in the late 1980s, in the 1990s, and in early 2000, interest in EOR and IOR flooding by the industry has re-emerged. As oil prices increased (at least up to mid-2008), EOR and IOR flooding and production became much more attractive. As the oil industry approaches worldwide peak production for conventional oil (it will happen, it is just a matter of when), the oil industry will have to begin to apply EOR and IOR on a large scale if conventional oil production is going to continue to make a major contribution to the world's energy demands and needs.

EOR flooding for application in conventional oil reservoirs encompasses two broad technology classes. The first EOR flooding type is *chemical EOR flooding*. Chemicals are primarily used in EOR flooding to reduce the interfacial tension in the reservoir between oil and water so that the residual-saturation oil can be recovered. Chemical EOR flooding processes include surfactant flooding, micellar flooding, micellar/polymer flooding, alkaline flooding, and alkaline/surfactant/polymer (ASP) flooding.

The second type of EOR flooding for conventional oil reservoirs is *miscible-gas flooding*. In this process, a gas is injected at sufficiently high pressures (pressure varies with different gases and different oils) to promote the miscibility of the mobile-gas phase in the oil phase (especially in the trapped-oil phase). Gases that can be used in EOR miscible-gas flooding include CO_2; hydrocarbon-miscible gas (e.g., C_2, C_3, and C_4); flue gas; and nitrogen. Miscible-gas flooding can be applied in the secondary and tertiary-recovery mode.

11.2 Problems of Earlier EOR Projects

Coming back to the contention made earlier in this book, if the target oil within a reservoir is not contacted by the oil-recovery drive fluid—in this case, by the EOR fluid—the target oil-recovery volume within the reservoir cannot be recovered by the EOR flood.

This can be visualized by studying **Fig. 11.1,** which is a reproduction of Fig. 2.1. The following discussion will be limited to chemical EOR flooding, but similar arguments hold also for gas EOR flooding. By viewing Fig. 11.1, it should be evident that if a limited volume of an expensive chemical EOR fluid is injected into the illustrated oil-reservoir conformance-problem situations, the chemical EOR fluid will tend to simply channel through the reservoir high-permeability channel. In doing so, the limited volume of the chemical EOR fluid will be highly ineffective in recovering oil from the extensive low-permeability portions of the reservoir. The volume of the conformance-problem high-permeability flow channels within oil reservoirs can be, and often are, quite small as compared to the total hydrocarbon pore volume. Thus, in this situation, the EOR flooding process will likely be deemed to be both a technical and an economic failure. Remember that EOR flooding can involve the recovery of both mobile and residual oil saturations residing in a reservoir at the beginning of the EOR project. *Underestimation of conformance problems and poor sweep efficiency limited the successful performance of many of the chemical and gas EOR field projects that were conducted in the 1970s and 1980s.*

For each of these failed EOR projects, specific technical issues were always reported to be the cause of most of the technical and economic failures of the EOR technologies applied. These technical issues included inadequate concentration of the chemicals used, inappropriate chemical flooding injection schemes, poor reservoir characterization, and injectivity loss due to the chemical injection, among others.

The performance of some of these early EOR flooding projects was later re-analyzed and re-visited. These analyses demonstrated that conformance problems were in fact the main cause that prevented the EOR chemicals fluids and gases from contacting nearly as much of the reservoir oil as expected. An example of this is the low-tension surfactant/polymer field test conducted in 1986 at the McCleskey sandstone formation

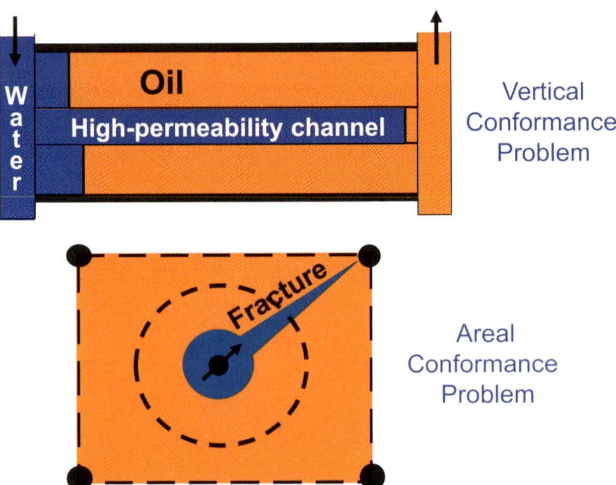

Fig.11.1—Conceptual illustration of oil-reservoir conformance problems.

in the Ranger oil field, Eastland County, Texas, USA, which, in 1992, Holly and Cayias re-evaluated, focusing on the design and operation of the test. Holly and Cayias (1992) determined that the main problem encountered was that the heterogeneity of the formation *was greater than expected.* This caused over injection of surfactant and polymer into high permeability zones. Although incremental oil was recovered, a better sweep efficiency of the target area would have increased oil production. Holly and Cayias (1992) concluded that the pilot test was successful on the basis of both monitored results and produced oil but that it did not achieve its design objectives because of reservoir heterogeneity and reduced sweep efficiency.

Another example is the polymer injection conducted in 1969 at the Ranger Zone Fault Block V in the Wilmington field located in Long Beach, California, USA. Krebs (1976) analyzed the failure of this polymer flooding, in which 1.3 million lbm of polymer, at an average concentration of 213 ppm, was injected with no significant increase in oil recovery. One of the main causes of the failure of this polymer flooding was early polymer breakthrough due to the existence of high permeability zones, which were not considered an important factor during the design of the polymer flooding. Additional factors that contributed to this polymer flooding failure were the use of a relatively low concentration of polymer and injectivity problems, which were caused by scale formation, undissolved polymer, and polymer adsorption. The poor results of this polymer flooding indicate the need for better reservoir information and improved knowledge of polymers and their behavior under reservoir conditions.

11.3 An Individual and Isolated Case

One of the authors of this book (Sydansk) was involved in extensive field-demonstration projects of a micellar-polymer EOR and residual-saturation oil recovery (RSOR) technology in the Illinois basin in the 1970s and 1980s. The micellar-polymer EOR technology was being applied and field-demonstrated in sandstone oil reservoirs that were characterized as being rather homogeneous in nature. A number of these field-demonstration projects did turn out to be moderately successful. The polymer component of the micellar-polymer flood is intended to promote better flooding sweep efficiency—both by improving mobility control and by improving sweep efficiency in the face of permeability heterogeneity.

In these field-demonstrations, it was detected that when a number of inter-well-pattern observation wells were drilled either during or after the micellar-polymer flooding, the displacement solution only swept small "high-permeability" vertical intervals of the reservoirs. The volume of the vertical intervals of these "homogeneous" sandstone oil reservoirs swept by the micellar-polymer solution was much smaller than the volume originally anticipated and much smaller than the volume required to promote a significant profitable performance. The project team clearly recognized that serious conformance problems were negatively impacting the performance of these Illinois-basin micellar-polymer EOR field-demonstration projects. At this time (mid-1980s), Sydansk was consulted and engaged in the development of conformance-improvement-treatment technologies that could help chemical EOR flooding to realize improved sweep efficiency and better economic performance, which resulted in the development of the Conformance-Improvement-Treatment Chromium(III) Gel (CIT) Technology in 1988 that has been widely applied in the field and proven to be successful.

11.4 EOR Conformance Problems—Planning Ahead

In view of growing interest in (in 2007 and early 2008) and the increasing need to exploit EOR and RSOR flooding in conventional oil reservoirs, it is important that *the underestimation of the conformance problems* that did negatively impact the technical and economic performance of many of EOR-flooding field-demonstration projects of the 1970s and 1980s be brought to the attention of operators, petroleum engineers, and oil-industry management in general who are now in the process of considering—or will be considering—the application of EOR-flooding techniques.

The recommended strategy during the design stage of any EOR and RSOR flooding operation is *to recognize and account for possible conformance problems and their significant negative impacts on the performance of the EOR-flooding operations. The design of the flooding operation should consider the appropriate risk factors and build in appropriate conformance-improvement and conformance-control techniques.* This approach should increase the technical and economical success of any EOR-flooding operation.

11.5 Conformance Improvement During EOR and RSOR Flooding

Conformance-improvement mobility-control components are incorporated into some RSOR flooding technologies—such as micellar-polymer and ASP-EOR flooding. Polymer has been historically incorporated into such RSOR flooding technologies (e.g., micellar-polymer flooding) primarily for the purpose of reducing viscous fingering of the drive fluid into and through the surfactant-EOR-fluid oil-recovery flood bank itself. In this case, a mini (chase) polymer flooding operation is conducted immediately after injecting the surfactant-containing RSOR fluid bank.

Alternatively, the polymer can be incorporated directly into the surfactant-EOR-driving fluid itself (e.g., as in ASP flooding) for the purpose of reducing viscous fingering of the surfactant-containing-EOR-fluid into the mobilized oil bank, where such fingering will lead to poor volumetric sweep efficiency within the well pattern, and thereby reduce oil recovery.

There seems to be an emerging consensus that polymer needs to be incorporated into *both* the primary EOR surfactant-containing displacement fluid, and the secondary drive (chasing) fluid bank that displaces the primary EOR surfactant-containing driving fluid. Additionally, consensus seems to be emerging in recognizing that the magnitude of the mobility-control problem should dictate how much polymer should be included in these two chemical-EOR-flooding fluid banks.

An alternative possible strategy for conformance improvement during a chemical EOR/RSOR flooding operation might be to increase the polymer concentration in the chemical surfactant flooding slug or to increase the size and/or the polymer concentration of the secondary polymer flood following the injection of a chemical surfactant slug. This would be done not only to improve the mobility control of the surfactant-solution flooding bank, but also to promote improved sweep efficiency and to generate additional incremental oil by recovering (by pure conformance-improvement considerations) unswept mobile oil saturation that is inevitably residing in the well pattern being flooded by the chemical EOR flood.

Nomenclature

C_m = mean interfacial curvature
E_A = areal sweep efficiency
E_I = vertical sweep efficiency
E_P = pattern sweep efficiency
E_V = volumetric sweep efficiency
h = net pay
h_f = film thickness
h_{fmax} = film thickness corresponding to Π_{max}
k = effective rock permeability
k_i = relative permeability to phase i (e.g., oil, brine, polymer solution)
M = mobility ratio
N_C = capillary number
P_C = capillary pressure
P_c^* = limiting capillary pressure
q = total injection or production rate
R_C = radius of curvature
r_e = external drainage radius
r_w = wellbore radius
t_1 = successive time 1
t_2 = successive time 2
t_3 = successive time 3
S_{or} = residual oil saturation
Δp = pressure drawdown of buildup
σ = surface tension
σ_{ow} = oil/water interfacial tension
γ_d = mobility of the displaced phase
γ_D = mobility of the oil-recovery displacing fluid
λ_d = mobility of the displaced fluid
λ_D = mobility of the oil-recovery displacing-fluid phase
λ_i = mobility of phase i
μ = viscosity
μ_i = viscosity of phase i
μ_w = viscosity of a flooding displacing fluid (usually water)
Π = film disjoining pressure
Π_{max} = maximum disjoining pressure
Π_{rup} = rupture disjoining pressure
ν = interstitial velocity

Abbreviations

ASP	Alkaline-surfactant-polymer
BOPD	Barrels of oil per day
CC/AP	Chromium(III)-carboxylate/acrylamide-polymer
CDG	Colloidal dispersion gel
EOR	Enhanced oil recovery
FWAG	Foam water-alternating-gas
HPAM	Hydrolyzed polyacrylamide
IOR	Improved oil recovery
OOIP	Original oil in place
PV	Pore volume
RPM	Relative permeability modification
WAG	Water-alternating-gas
RSOR	Residual-saturation oil recovery
WOR	Water/oil ratio

Glossary

Advanced oil recovery: Includes all oil-recovery techniques that are more technically sophisticated than conventional recovery methods (i.e., than standard primary and secondary recovery methods).

Areal conformance: A measure of the uniformity of the propagation of the flood front of the injected drive fluid in the areal x-y plane during an oil-recovery flooding operation.

Areal conformance problems: Directional conformance problems that cause uneven and poor oil-recovery drive-fluid sweep in the areal x-y plane of a petroleum reservoir.

Bulk gel: A classical gel where the internal chemical gel structure is nearly infinite, and on the macro scale, the gel structure is continuous throughout the gel mass.

Capillary number: For waterflooding, is mathematically defined as the ratio of viscous forces divided by the capillary forces operative on an oil drop in a water-wet pore.

CDG: Colloidal dispersion gels are microgels formulated with 150–1,200 ppm (below the polymer's critical overlap concentration) (e.g., high-molecular-weight hydrolyzed polyacrylamide that is crosslinked either by aluminum citrate or chromic triacetate).

Conformance: A measure of oil-recovery drive-fluid sweep efficiency.

Conformance improvement technique: Any technique or operation that is applied to reduce or eliminate an oil-recovery conformance problem.

Conformance improvement treatment: A permeability-reducing treatment, usually of small volume (e.g., 10 to several 10,000 bbl), that is placed within a reservoir to reduce or eliminate an oil-recovery conformance problem.

Conformance problem: A problem involving inefficient oil recovery or less-than-ideal oil recovery, which results from spatial variation in permeability within the reservoir; such problems include, by definition, excessive-water-production problems.

Conventional oil recovery: Includes all customary oil-recovery techniques, specifically primary and secondary oil recovery, but specifically excludes advanced oil recovery.

Conventional oil reservoir: An oil reservoir in which the viscosity of the oil is less than 200 cp and/or the API gravity is greater than 20.

Critical overlap concentration: For polymer dissolved in a solvent, the polymer concentration below which polymer molecules tumble in solution; the polymer molecules do not directly interact with and contact one another.

Crossflow: The vertical flow of fluids between reservoir strata of differing permeability that overlay one another.

Cusping conformance problem: The undesirable flow of water up, or of gas down, into a wellbore through an inclined high-permeability reservoir strata.

Disproportionate permeability reduction (DPR): A process in which the permeability of one fluid (normally water) is reduced to a greater extent than the permeability of other reservoir fluids (normally oil and gas).

Economic oil-recovery efficiency: The volume of oil recovered per cost of the oil-recovery process, or (as used in the context of elapsed time) the volume of oil recovered as a function of elapsed recovery time.

Enhanced oil recovery (EOR): Enhanced oil recovery is promoted by the injection of materials not normally present in an oil reservoir.

Gel: A fluid-based system to which at least some solid-like structural properties have been imparted.

Gelant: The pregel solution of a gel; the pregel solution before any chemical reaction of gel formation has occurred (e.g., before chemical crosslinking of polymers).

High permeability anomaly: Within a petroleum reservoir, a flow channel having exceptionally high permeability (>10 darcies), such as fractures, solution channels, or interconnected vugular porosity.

Imbibition: Absorption of water by a water-wet matrix-reservoir-porous media.

Immobile oil: The fraction of the oil saturation within a reservoir that is not mobilized and recovered by a waterflood (or by another immiscible flooding fluid). It is oil saturation trapped in the porous media by capillary pressure.

Improved oil recovery (IOR): Improved oil recovery is incremental oil recovery that results from the application of any advanced oil-recovery technique that is applied during any type of ongoing oil-recovery operation.

Incremental oil recovery: Oil recovery that results from applying advanced, improved, or enhanced oil-recovery techniques. It is the oil recovery obtained beyond what would have been recovered otherwise.

Matrix rock: The bulk of petroleum reservoir rock that has "normal" permeabilities (<10 darcies, and most often <1 darcy) and does not contain any high-permeability-anomaly (>10 darcy) flow channels.

Matrix-rock reservoirs: A reservoir that is comprised solely of matrix rock and does not contain high-permeability anomalies (such as natural fractures).

Microgel: A relatively low-polymer-concentration gel that forms discrete microscale gel particles within a liquid solution (e.g., water).

Mobile oil: Oil that can be mobilized in, and recovered from, reservoir matrix rock during flooding with an immiscible fluid (e.g., water).

Mobility: The permeability to a given fluid divided by the viscosity of that fluid as it flows through reservoir matrix rock at a given fluid saturation condition.

Mobility ratio: The mobility of the oil-recovery flooding displacing fluid (e.g., water) divided by the mobility of the displaced oil phase during an oil-recovery flooding operation.

Original oil in place (OOIP): Original oil in place is all of the original oil in place within an oil reservoir—including the associated residual oil saturation.

Primary oil recovery: Oil production that results either from the natural energy and forces of the reservoir (naturally flowing wells) or with only the aid of artificial-lift production; primary oil recovery does not involve any flooding operation. (It is often referred to as the first "crop" of oil production.)

Profile modification treatment: A near-wellbore permeability-reducing conformance-improvement treatment applied in a matrix-rock reservoir suffering from vertical conformance problems that result from geological strata of differing permeability overlaying one another.

Relative permeability modification: A process or treatment in which the water/oil relative-permeability curve is altered. It refers to the use of a material (e.g., a water-shutoff treatment material) that reduces the permeability to water flow to a greater extent than the permeability to oil or gas flow.

Residual oil saturation: The oil saturation in reservoir matrix rock that cannot be recovered by waterflooding—even in the presence of perfect flood conformance and/or flood sweep efficiency.

Residual-saturation oil recovery (RSOR): Oil-recovery processes and technologies that are primarily intended for recovering immobile residual oil saturation.

Secondary oil recovery: The addition of recovery energy and forces to produce oil beyond what can be produced by primary recovery—examples are waterflooding and pressure maintenance operations. (Secondary oil recovery is often referred to as the second "crop" of oil production).

Sweep efficiency: A measure of the degree of the uniform expansion of the flooding-fluid front during an oil-recovery flooding operation.

Tertiary oil recovery: Oil recovery beyond primary and secondary recovery; it normally involves augmented-oil-flow or augmented-oil-recovery agents being added to a flooding fluid.

Vertical conformance: A measure of the uniformity of the propagation of the flood front of the injected drive fluid in the vertical z-x plane during an oil-recovery flooding operation.

Vertical conformance problem: Conformance problems causing uneven and poor oil-recovery drive-fluid sweep in the vertical direction in a petroleum reservoir.

Viscous fingering: Fluid fingering of the oil-displacing fluid (e.g., water) into the fluid (e.g., oil) that is being displaced during an oil-recovery flooding operation, in which the fingering is caused by a poor mobility ratio and/or a low-viscosity fluid displacing a high-viscosity fluid.

Volumetric sweep efficiency: The sweep efficiency of an oil-recovery flood in the combined vertical and areal directions within a petroleum reservoir.

Vug: Void spaces found in matrix (usually carbonate) reservoirs where the diameters of such voids range from about 1 mm up to several centimeters.

References

Abbasy, I., Vasquez, J., Eoff, L., and Dalrymple, D. 2008. Laboratory Evaluation of Water-Swellable Materials for Fracture Shutoff. Paper SPE 113193 presented at the SPE/DOE Improved Oil Recovery Symposium, Tulsa, 20–23 April. DOI: 10.2118/113193-MA.

Alhajeri, M.M., Green, D.W., Liang, J., and Pancake, R.E. 2006. Gel-Polymer Extends Arbuckle High-Water-Cut Well Life. *Oil & Gas Journal* **104** (2): 39–43.

Anderson, D.M., Stotts, G.W.J., Mattar, L., Ilk, D., and Blasingame, T.A. 2006. Production Data Analysis-Challenges, Pitfalls, Diagnostics. Paper SPE 102048 presented at the SPE Annual Technical Conference and Exhibition, San Antonio, Texas, USA, 24–27 September. DOI: 10.2118/102048-MS.

Apydin, O., Bertin, H., Catanier, L., Kovscek, A. 1998. An Experimental Investigation of Foam Flow in Homogeneous and Heterogeneous Porous Media. SUPRI TR-112 Report, Contract No. DE-FG22-96BC14994, US DOE, Washington, DC (June 1998).

Asadi, M. and Shook, M. 2010. Application of Chemical Tracers in IOR: A Case History. Paper SPE 126029 presented at the North Africa Technical Conference and Exhibition, Cairo, Egypt, 14–17 February. DOI: 10.2118/126029-MS.

Bai, B., Huang, F., Liu, Y., Seright, R.S., and Wang, Y. 2008. Case Study on Preformed Particle Gel For In-Depth Fluid Diversion. Paper SPE 113997 presented at the SPE/DOE Symposium on Improved Oil Recovery, Tulsa, 20–23 April. DOI: 10.2118/113997-MS.

Bai, B., Li, L., Liu, Y., Liu, H., Wang, Z., and You, C. 2007. Preformed Particle Gel for Conformance Control: Factors Affecting Its Properties and Applications. *SPE Res Eval & Eng* **10** (4): 415–422. SPE-89389-PA. DOI: 10.2118/89389-PA.

Bensted, J. and Barnes, P. eds. 2002. *Structure and Performance of Cements,* 247–248. New York City: Taylor & Francis Group.

Blount, C.G., Mooney, M.B., Behenna, F.R., Stephens, R.K., and Smith, R.D. 2006. Well-Intervention Challenges to Service Wells that Can Be Drilled. Paper 100172 presented at the SPE/ICoTA Coiled Tubing Conference and Exhibition, The Woodlands, Texas, USA, 4–5 April. DOI: 10.2118/100172-MS.

Boisnault, J.M., Bourahla, A., Tirlia, T., Holmes, C., Raiturkar, A.M., Maroy, P., Moffett, C., Perez, G., Ramirez, I., Revil, P., and Roemer, R. 1999. Concrete Developments in Cementing Technology. *Oilfield Review* **11** (1):18–29.

Borling, D.C. 1994. Injection Conformance Control Case Histories Using Gels at the Wertz Field CO_2 Tertiary Flood in Wyoming. Paper SPE 27825 presented at the SPE/DOE Symposium on Improved Oil Recovery, Tulsa, 17–20 April. DOI: 10.2118/27825-MS.

Calhoun, T.G. and Hurford, G. 1970. Case History of Radioactive Tracers and Techniques in Fairway Field. *J Pet Technol* **22** (10): 1217–1224. SPE-2853-PA. DOI: 10.2118/2853-PA.

Chambers, K. and Radke, J. 1991. Capillary Phenomena in Foam Flow Through Porous Media. In *Interfacial Phenomena in Petroleum Recovery,* ed. N. Morrow, Chap. 6, 191–255. New York City: Surfactant Science Series, Marcel Dekker.

Chang, H.L., Zhang, Z.Q., Wang, Q.M., Xu, Z.S., Guo, Z.D., Sun, H.Q., Cao, X.L., et al. 2006. Advances in Polymer Flooding and Alkaline/Surfactant/Polymer Processes as Developed and Applied in the People's Republic of China. *J Pet Technol* **58** (2): 84–89. SPE-89175-MS. DOI : 10.2118/89175-MS

Chauveteau, G., Omari, A., Tabary R., Renard, M., Veerapen, J., and Rose, J. 2001. New Sized-Controlled Microgels for Oil Production. Paper SPE 64988 presented at the SPE International Symposium on Oilfield Chemistry, Houston, 13–16 February. DOI: 10.2118/64988-MS.

Chauveteau, G., Tabary, R., le Bon, C., Renard, M., Feng, Y., and Omari, A. 2003. In-Depth Permeability Control by Adsorption of Soft Sized-Controlled Microgels. Paper SPE 82228 presented at the SPE European Formation Damage Conference, The Hague, 13–14 May. DOI: 10.2118/82228-MS.

Chou, S.I., Bae, J.H., Friedmann, F., and Dolan, J.D. 1994. Development of Optimal Water Control Strategies. Paper SPE 28571 presented at the SPE Annual Technical Conference and Exhibition, New Orleans, 25–28 September. DOI: 10.2118/28571-MS.

Christopher, C.A., Clark, T.J., and Gibson, D.H. 1988. Performance and Operation of a Succesful Polymer Flood in the Sleepy Hollow Reagan Unit. Paper SPE 17395 presented at the SPE/DOE Enhanced Oil Recovery Symposium, Tulsa, 16–21 April. DOI: 10.2118/17395-MS.

Clarke, W.J. and McNally, A.C. 1993. Ultrafine Cement for Oilwell Cementing. Paper SPE 25868 presented at the SPE Low Permeability Reservoirs Symposium, Denver, 12–14 April. DOI: 10.2118/25868-MS.

Cohen, D., Patzek, T.W., and Radke, C.J. 1996. Two-Dimensional Network Simulation of Diffusion-Driven Coarsening of Foam Inside a Porous Medium. *J. of Colloid and Interface Science* **179** (2): 357–373.

Cossé, R. 1993. *Basic Reservoir Engineering,* 272. Paris: Technip Editions.

Craig, F.G. 1971. *The Reservoir Engineering Aspects of Waterflooding.* Monograph Series, SPE, Richardson, Texas **3:** 48–75.

Danardatu, H., Martin, D., and Colpitts, R. 2005. Reservoir Engineering. In *Standard Handbook of Petroleum & Natural Gas Engineering,* ed. W. Lyons and G. Plisga, Chap. 5, 5-224–5-225. Burlington, Massachusetts: Elsevier.

Delshad, M., Kim, D.H., Magbagbeola, O.A., Huh, C., Pope, G.A., and Tarahhom, F. 2008. Mechanistic Interpretation and Utilization of Viscoelastic Behavior of Polymer Solutions for Improved Polymer-Flooding Efficiency. 2008. Paper SPE 113620 presented at the SPE/DOE Symposium on Improved Oil Recovery, Tulsa, 20–23 April. DOI: 10.2118/113620-MS.

Demin, W., Jiecheng, C., Junzheng, W., and Gang, W. 2002a. Experiences Learned After Production of More than 300 Million Barrels of Oil by Polymer Flooding in

Daqing Oil Field. Paper SPE 77693 presented at the SPE Annual Technical Conference and Exhibition, San Antonio, Texas, USA, 29 September–2 October. DOI: 10.2118/77693-MS.

Demin, W., Jiecheng, C., Qingyan, Y., Wenchao, G., Qun, L., and Fuming, C. 2000. Viscous-Elastic Polymer Can Increase Microscale Displacement Efficiency in Cores. Paper SPE 63227 presented at the SPE Annual Technical Conference and Exhibition, Dallas, 1–4 October. DOI: 10.2118/63227-MS.

Demin, W., Yingjie, S., Yan, W., and Xuping, T. 2002b. Producing More Than 75% of Daqing Oil Field's Production by IOR, What Experiences Have Been Learnt? Paper SPE 77871 presented at the SPE Asia Pacific Oil and Gas Conference and Exhibition, Melbourne, Australia, 8–10 October. DOI: 10.2118/77871-MS.

Demin, W., Youlin, J., Yan, W., Xiaohong, G., and Gang, W. 2004. Viscous-Elastic Polymer Fluids Rheology and Its Effect Upon Production Equipment. *SPE Prod & Fac* **19** (4): 209–216. SPE-77496-PA. DOI: 10.2118/77496-PA.

Djabbarah, N.F., Weber, S.L., Freeman, D.C., Muscatello, J.A., Ashbaugh, J.P., and Covington, T.E. 1990. Laboratory Design and Field Demonstration of Steam Diversion With Foam. Paper SPE 20067 presented at the SPE California Regional Meeting, Ventura, California, USA, 4–6 April. DOI: 10.2118/20067-MS.

Dupuis, G., Rousseau, D., Tabary, R., and Grassl, B. 2010. How to Get the Best Out of Hydrophobically Associative Polymers for IOR? New Experimental Insights. Paper SPE 129884 presented at the SPE Improved Oil Recovery Symposium, Tulsa, 24–28 April. DOI: 10.2118/129884-MS.

Dusseault, M.B. 2007. Cold Heavy-Oil Production With Sand. In *Petroleum Engineering Handbook,* ed. L.W. Lake, Vol. VI, Chap. 6, 183. Richardson, Texas: SPE.

Feng, J., Zhao, J., Chen, S., and Qu, C. 1994. Water/Oil Separation Characteristics of Daqing Oilfield Polymer Flooding Production Fluid. Paper SPE 28540 presented at the SPE Annual Technical Conference and Exhibition, New Orleans, 25–28 September. DOI: 10.2118/28540-MS.

Fletcher, A.J. and Davis, J.P. 2010. How EOR Can Be Transformed by Nanotechnology. Paper SPE 129531 presented at the SPE Improved Oil Recovery Symposium, Tulsa, 24–28 April. DOI: 10.2118/129531-MS.

Flores, J.G., Elphick, J., Lopez, F., and Espinel, P. 2008. The Integrated Approach to Formation Water Management: From Reservoir Management to Protection of the Environment. Paper SPE 116218 presented at the SPE Annual Technical Convention and Exhibition, Denver, 21–24 September. DOI: 10.2118/116218-MS.

Frampton, H., Denyer, P., Ohms, D., Husband, M., and Mustoni, J. 2009. Sweep Improvement from the Lab to the Field. Paper B22 presented at the European Symposium on Improved Oil Recovery, Paris, 27–29 April.

Frampton, H., Morgan, J.C., Cheung, S.K., Munson, L., Chang, K.T., and Williams, D. 2004. Development of a Novel Waterflood Conformance Control System. Paper SPE 89391 presented at the SPE/DOE Symposium on Improved Oil Recovery, Tulsa, 17–21 April. DOI: 10.2118/89391-MS.

Friedmann, F., Hughes, T.L., Smith, M.E., Hild, G.P., Wilson, A., and Davies, S.N. 1999. Development and Testing of a Foam-Gel Technology to Improve Conformance of the Rangely CO_2 Flood. *SPE Res Eval & Eng* **2** (1): 4–13. SPE-54429-PA. DOI: 10.2118/54429-PA.

Geffen, T.M. 1973. Improved Oil Recovery Could Help Ease Energy Shortage. *World Oil* **177** (October 1973): 84–88.

Green, D.W. and Willhite, G.P. 1998. *Enhanced Oil Recovery*. Textbook Series, SPE, Richardson, Texas **6:** 545.

Hanssen, J.E. and Dalland, M. 1994. Gas Blocking Foams. In *Foams: Fundamentals and Applications in the Petroleum Industry,* ed. L.L. Schramm, Chap. 8, 319–353. Washington, DC: Advances in Chemistry Series 242, American Chemical Society.

Heller, J.P. 1994. CO_2 Foams in Enhanced Oil Recovery. In *Foams: Fundamentals and Applications in the Petroleum Industry,* ed. L.L. Schramm, Chap. 5, 201-234. Washington, DC: Advances in Chemistry Series 242, American Chemical Society.

Hirasaki, G.J. 1989. The Steam-Foam Process. *J Pet Technol* **41** (4): 449–456. SPE-19505-PA. DOI: 10.2118/19505-PA.

Hirasaki, G.J., Miller, C.A., and Puerto, M. 2008. Recent Advances in Surfactant EOR. Paper SPE 115386 presented at the SPE Annual Technical Conference and Exhibition, Denver, 21–24 September. DOI: 10.2118/115386-MS.

Hild, G.P. and Wackowski, R.K. 1999. Reservoir Polymer Gel Treatments to Improve Miscible CO_2 Flood. *SPE Res Eval & Eng* **2** (2): 196–204. SPE-56008-PA. DOI: 10.2118/56008-PA.

Holley, S.M. and Cayias, J.L. 1992. Design, Operation, and Evaluation of a Surfactant/Polymer Field Pilot Test. *SPE Res Eng* **7** (1): 9–14. SPE-20232-PA. DOI: 10.2118/20232-PA.

Hughes, T.L., Friedmann, F., Johnson, D., Hild, G.P., Wilson, A., and Davies, S.N. 1999. Large-Volume Foam-Gel Treatments to Improve Conformance of the Rangely CO_2 Flood. *SPE Res Eval & Eng* **2** (1): 14–24. SPE-54772-PA. DOI: 10.2118/54772-PA.

Huh, C. and Rossen, W.R. 2008. Approximate Pore-Level Modeling for Apparent Viscosity of Polymer-Enhanced Foam in Porous Media. *SPE J.* **13** (1): 17–25. SPE-99653-PA. DOI: 10.2118/99653-PA.

Jaripatke, O. and Dalrymple, D. 2010. Water-Control Management Technologies: A Review of Successful Chemical Technologies in the Last Two Decades. Paper SPE 127806 presented at the SPE International Symposium and Exhibition on Formation Damage Control, Lafayette, Louisiana, USA, 10–12 February. DOI: 10.2118/127806-MS.

Jiang, H., Wu, W., Wang, D., Zeng, Y., Zhao, S., and Nie, J. 2008. The Effect of Elasticity on Displacement Efficiency in the Lab and Results of High Concentration Polymer Flooding in the Field. Paper SPE 115315 presented at the SPE Annual Technical Conference and Exhibition, Denver, 21–24 September. DOI: 10.2118/115315-MS.

Kovscek, A.R. and Radke, C.J. 1994. Fundamentals of Foam Transport in Porous Media. In *Foams: Fundamentals and Applications in the Petroleum Industry,* ed. L.L. Schramm, Chap. 3, 115–163. Washington, DC: Advances in Chemistry Series 242, American Chemical Society.

Krebs, H.J. 1976. The Wilmington Field, California, Polymer Flood A Case History. *J Pet Technol* **28** (12):1473–1480. SPE-5828-PA. DOI: 10.2118/5828-PA.

Laherrere, J. 2003. Future of Oil Supplies. *Energy Exploration & Exploitation* **21** (3): 227–267.

Lakatos, I. and Lakatos-Szabó, J. 2008. Global Oil Demand and the Role of Chemical EOR Methods in the 21st Century. *Int. J. of Oil, Gas, and Coal Technology* **1** (1/2): 46–64.

Lake, L.W. 1989. *Enhanced Oil Recovery.* Englewood Cliffs, New Jersey: Prentice Hall.

Lane, R.H. and Sanders, G.S. 1995. Water Shutoff through Fullbore Placement of Polymer Gel in Faulted and in Hydraulically Fractured Producers of the Prudhoe Bay Field. Paper SPE 29475 presented at the SPE Production Operations Symposium, Oklahoma City, Oklahoma, USA, 2–4 April. DOI: 10.2118/29475-MS.

Liang, J.T., Lee, R.L., and Seright, R.S. 1993. Gel Placement in Production Wells. *SPE Prod & Fac* **8** (4): 276–284. SPE-20211-PA. DOI: 10.2118/20211-PA.

Liu, Y., Bai, B., and Shuler, P.J. 2006. Application and Development of Chemical-Based Conformance Control Treatments in China Oil Fields. Paper SPE 99641 presented at the SPE/DOE Symposium on Improved Oil Recovery, Tulsa, 22–26 April. DOI: 10.2118/99641-MS.

Mack, J.C. and Smith, J.E. 1994. In-Depth Colloidal Dispersion Gels Improve Oil Recovery Efficiency. Paper SPE 27780 presented at the SPE/DOE Symposium on Improved Oil Recovery, Tulsa, 17–20 April. DOI: 10.2118/27780-MS.

Miglicco, T.P. 1986. Polymer Flood Operation: East Texas Field. Paper SPE 14658 presented at the SPE East Texas Regional Meeting, Tyler, Texas, USA, 21–22 April. DOI: 10.2118/14658-MS.

Moffitt, P.D. 1993. Long-Term Production Results of Polymer Treatments in Producing Wells in Western Kansas. *J Pet Technol* **45** (4): 356–362. SPE-22649-PA. DOI: 10.2118/22649-PA.

Nguyen, Q., Alexandrov, A., Zitha, P., and Currie, P. 2000. Experimental and Modeling Studies on Foam in Porous Media: A Review. Paper SPE 58799 presented at the SPE International Symposium on Formation Damage Control, Lafayette, Louisiana, USA, 23–24 February. DOI: 10.2118/58799-MS.

Ohms, D., McLeod, J., Graff, C.J., Frampton, H., Morgan, J.C., Cheung, S., Yancey, K., and Chang, K.T. 2009. Incremental Oil Success From Waterflood Sweep Improvement in Alaska. Paper SPE 121761 presented at the SPE International Symposium on Oilfield Chemistry, The Woodlands, Texas, USA, 20–22 April. DOI: 10.2118/121761-MS.

Petzet, A., Moritis, G., Fletcher, S., Dittrick, P., and Izundu, U. 2009. 40th OTC Highlights World Economy, Energy Policy, and Environment. *Oil and Gas Journal* **107** (18): 20–27.

Portwood, J.T. 2005. The Kansas Arbuckle Formation: Performance Evaluation and Lessons Learned from More Than 200 Polymer-Gel Water-Shutoff Treatments. Paper SPE 94096 presented at the SPE Production Operations Symposium, Oklahoma City, Oklahoma, USA, 16–19 April. DOI: 10.2118/94096-MS.

Pritchett, J., Frampton, H., Brinkman, J., Cheung, S., Morgan, J., Chang, K.T., Williams, D., et al. 2003. Field Application of a New In-Depth Waterflood Conformance Improvement Tool. Paper 84897 presented at the SPE International Improved Oil Recovery Conference in Asia Pacific, Kuala Lumpur, 20–12 October. DOI: 10.2118/84897-MS.

Reynolds, R.R. 2003. Gel Polymer Treatments in Kansas Arbuckle Wells. http://www.nmcpttc.org/Case_Studies/GelPolymer/index.html (accessed 16 January 2009).

Romero-Zeron, L. 2004. Wettability Effect on Foamed Gel Flow. PhD disserataion, University of Calgary, Calgary, Alberta, Canada.

Rossen, W.R. 1996. Foams in Enhanced Oil Recovery. In *Foams–Theory, Measurement, and Applications,* ed. R.K. Prud'homme and S.A. Khan, Chap. 11, 413–464. New York: Marcel Dekker.

Sandrea, I. and Sandrea, R. 2007. Global Oil Reserves–1: Recovery Factors Leave Vast Target for EOR Technologies. *Oil & Gas Journal* **105** (41): 44–47.

Satter, A., Iqbal, G.M., and Buchwalter, J. 2008. *Practical Enhanced Reservoir Engineering,* 504–505. Tulsa: PennWell.

Schramm, L.L. 1994. Foam Sensitivity to Crude Oil in Porous Media. In *Foams: Fundamentals and Applications in the Petroleum Industry,* ed. L.L. Schramm, Chap. 4, 165–197. Washington, DC: Advances in Chemistry Series 242, American Chemical Society.

Schramm, L.L. and Wassmuth, F. 1994. Foams: Basic Principles. In *Foams: Fundamentals and Applications in the Petroleum Industry,* ed. L.L. Schramm, Chap. 1, 3–45. Washington, DC: Advances in Chemistry Series 242, American Chemical Society.

Seright, R.S. 1988. Placement of Gels To Modify Injection Profiles. Paper SPE 17332 presented at the SPE Enhanced Oil Recovery Symposium, Tulsa, 16–21 April. DOI: 10.2118/17332-MS.

Seright, R.S. 2001. Gel Propagation Through Fractures. *SPE Prod & Fac* **16** (4): 225–231. SPE-74602-PA. DOI: 10.2118/74602-PA.

Seright, R.S. 2006. Discussion of SPE 89175, Advances in Polymer Flooding and Alkaline/Surfactant/Polymer Processes as Developed and Applied in the People's Republic of China. *J Pet Technol* **58** (2): 80. SPE-89175-MS. DOI: 10.2118/89175-MS.

Seright, R.S. 2010. Potential for Polymer Flooding Reservoirs with Viscous Oils. Paper SPE 129899 presented at the SPE Improved Oil Recovery Symposium, Tulsa, 24–28 April. DOI: 10.2118/129899-MS.

Seright, R.S. and Liang, J. 1995. A Comparison of Different Types of Blocking Agents. Paper SPE 30120 presented at the SPE European Formation Damage Conference, The Hague, 15–16 May. DOI: 10.2118/30120-MS.

Seright, R.S., Lane, R.H., and Sydansk, R.D. 2003. A Strategy for Attacking Excess Water Production. *SPE Prod & Fac* **18** (3): 158–169. SPE-84966-PA. DOI: 10.2118/84966-PA.

Seright, R.S., Prodanovic, M., and Lindquist, W.B. 2006. X-Ray Computed Microtomography Studies of Fluid Partitioning in Drainage and Imbibition Before and After Gel Placement. *SPE J.* **11** (2): 159–170. SPE-89393-PA. DOI: 10.2118/89393-PA.

Seright, R.S., Seheult, M., Talashek, T. 2009. Injectivity Characteristics of EOR Polymers. *SPE Res Eval & Eng* **12** (5): 783–792. SPE-115142-PA. DOI: 10.2118/115142-PA.

Shen, C., Nguyen, Q., Huh, C., and Rossen, W.R. 2006. Does Polymer Stabilize Foam in Porous Media? Paper SPE 99796 presented at the SPE/DOE Symposium on Improved Oil Recovery, Tulsa, 22–26 April. DOI: 10.2118/99796-MS.

Shirley, A.I. 1988. Foam Formation in Porous Media: A Microscopic Visual Study. In *Surfactant-Based Mobility Control,* ed. D. Smith, Chap. 12, 234–257. Washington, DC.: American Chemical Society Symposium Series.

Smith, J.E. 1995. Performance of 18 Polymers in Aluminum Citrate Colloidal Dispersion Gels. Paper 28989 presented at the SPE International Symposium on Oilfield Chemistry, San Antonio, Texas, USA, 14–17 February. DOI: 10.2118/28989-MS.

Smith, R.J. and Perepelecta, K.R. 2002. Steam Conformance Along Horizontal Wells at Cold Lake. Paper SPE 79009 presented at the SPE International Thermal Operations and Heavy Oil Symposium and International Horizontal Well Technology Conference, Calgary, Alberta, Canada, 4–7 November. DOI: 10.2118/79009-MS.

Soliman, M.Y., Creel, P., Rester, S., Sigal, R., Everett, D., and Johnson, M.H. 2000. Integration of Technology Supports Preventive Conformance Reservoir Techniques. Paper SPE 62553 presented at the SPE/AAPG Western Regional Meeting, Long Beach, California, USA, 19–23 June. DOI: 10.2118/62553-MS.

Sorbie, K.S. 1991. *Polymer-Improved Oil Recovery,* 376. Glasgow: Blackie.

Sorbie, K.S. and Seright, R.S. 1992. Gel Placement in Heterogeneous Systems with Crossflow. Paper SPE 24192 presented at the SPE/DOE Enhanced Oil Recovery Symposim, Tulsa, 22–24 April. DOI: 10.2118/24192-MS.

Spildo, K., Skauge, A., Aarra, M.G., and Tweheyo, M.T. 2008. A New Polymer Application for North Sea Reservoirs. Paper SPE 113460 presented at the SPE/DOE Symposium on Improved Oil Recovery, Tulsa, 20–23 April. DOI: 10.2118/113460-MS.

Stosur, G.J, Hite, J.R., Carnahan, N.F., and Miller, K. 2003. The Alphabet Soup of IOR, EOR and AOR: Effective Communication Requires a Definition of Terms. Paper 84908 presented at the SPE International Improved Oil Recovery Conference in Asia Pacific, Kuala Lumpur, 20–21 October. DOI: 10.2118/84908-MS.

Sydansk, R.D. 1990. A Newly Developed Chromium(III) Gel Technology. *SPE Res Eng* **5** (3): 346–352. SPE-19308-PA. DOI: 10.2118/19308-PA.

Sydansk, R.D. 1993. Acrylamide-Polymer/Chromium(III)-Carboxylate Gels for Near Wellbore Matrix Treatments. *SPE Advanced Technology Series* **1** (1): 146–152. SPE-20214-PA. DOI: 10.2118/20214-PA.

Sydansk, R.D. 2007. Polymers, Gels, Foams, and Resins. In *Petroleum Engineering Handbook,* ed. L.W. Lake, Vol. V(B), Chap. 13, 1149–1260. Richardson, Texas: SPE.

Sydansk, R.D. and Moore, P.E. 1990. Production Responses in Wyoming's Big Horn Basin Resulting From Application of Acrylalmide-Polymer/CrIII-Carboxylate Gels. Paper SPE 21894 available from SPE, Richardson, Texas.

Sydansk, R.D. and Seright, R.S. 2007. When and Where Relative Permeability Modification Water-Shutoff Treatments Can Be Successfully Applied. *SPE Prod & Oper* **22** (2): 236–247. SPE-99371-PA. DOI: 10.2118/99371-PA.

Sydansk, R.D. and Southwell, G.P. 2000. More Than 12 Years' Experience With a Successful Conformance-Control Polymer-Gel Technology. *SPE Prod & Oper* **15** (4): 270–278. SPE-66558-PA. DOI: 10.2118/66558-PA.

Talukdar, S. and Instefjord, R. 2008. Reservoir Management of the Gullfaks Main Field. Paper SPE 113260 presented at the Europec/EAGE Conference and Exhibition, Rome, 9–12 June. DOI: 10.2118/113260-MS.

Tanzil, D., Hirasaki, G.J., and Miller, C.A. 2002. Mobility of Foam in Heterogeneous Media: Flow Parallel and Perpendicular to Stratification. *SPE J.* **7** (2): 203–212. SPE-78601-PA. DOI: 10.2118/78601-PA.

Taylor, K. and Hawkins, B. 1992. Emulsions in Enhanced Oil Recovery. In *Emulsions: Fundamentals and Applications in the Petroleum Industry,* ed. L. Schramm,

Advances in Chemistry, American Chemical Society, **231**, Chap. 7, 263–293. Oxford University Press.

Veil, J.A., Puder, M.G., Elcock, D., and Redweik, R.J. 2004. A White Paper Describing Produced Water from Production of Crude Oil, Natural Gas, and Coal Bed Methane. Argonne National Laboratory. Final Report, Contract W-31-109-Eng-38, US DOE NETL, Argonne, Illinois (January 2004).

Wang, D., Han, P., Shao, Z., Hou, W. and Seright, R.S., 2008a. Sweep Improvement Options for the Daqing Oil Field. *SPE Res Eval & Eng* **11** (1): 18–26. SPE-99441-PA. DOI: 10.2118/99441-PA.

Wang, D., Seright, R.S., Shao, Z., and Wang, J. 2008b. Key Aspects of Project Design for Polymer Flooding at the Daqing Oil Field. *SPE Res Eval & Eng* **11** (6): 1117–1124. SPE-109682-PA. DOI: 10.2118/109682-PA.

Willhite, G.P. and Pancake, R.E. 2008. Controlling Water Production Using Gelled Polymer Systems. *SPE Res Eval & Eng* **11** (3): 454–465. SPE-89464-PA. DOI: 10.2118/89464-PA.

Willhite, G.P., Zhu, H., Natarajan, D., McCool, C.S., and Green, D.W. 2002. Mechanisms Causing Disproportionate Permeability Reduction in Porous Media Treated With Chromium Acetate/HPAM Gels. *SPE J.* **7** (1):100–108. SPE-77185-PA. DOI: 10.2118/77185-PA.

Yan, W., Demin, W., Jun, W., Jiangtao, L., Runtao, Y., and Zengyou, D. 2008. New Developments in Production Technology for Polymer Flooding. Paper SPE 114336 presented at the SPE/DOE Improved Oil Recovery Symposium, Tulsa, 19–23 April. DOI: 10.2118/114336-MS.

SI Metric Conversion Factors

°API		$141.5/(131.5 + °API)$		$= g/cm^3$
acre-ft	×	1.233 489	E + 03	$= m^3$
bbl	×	1.589 873	E – 01	$= m^3$
Ci	×	3.7*	E + 10	= Bq
cp	×	1.0*	E – 03	= Pa·s
ft	×	3.048*	E – 01	= m
°F		$(°F – 32)/1.8$		= °C
in.	×	2.54*	E + 00	= cm
lbm	×	4.535 924	E – 01	= kg

*Conversion factor is exact.

Author Index

Subject Index